ROBOTS, AN INTRODUCTION TO BAS
TS 191.8 O78 1983
A05441651001

I0429736

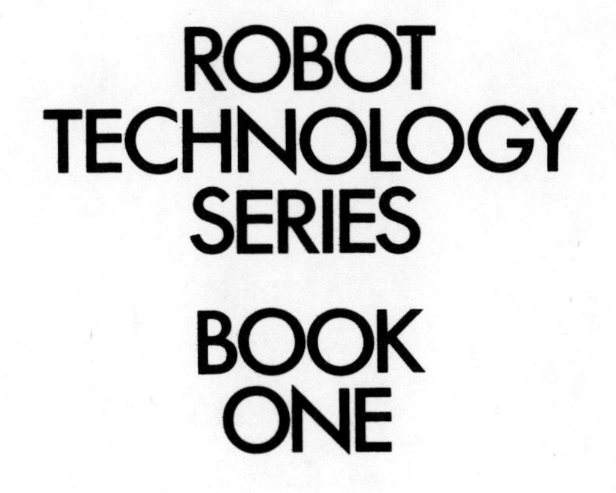
ROBOT
TECHNOLOGY
SERIES
BOOK
ONE

# ROBOTS

## AN INTRODUCTION TO BASIC CONCEPTS AND APPLICATIONS

BY DAVID M. OSBORNE

by Midwest Sci-Tech Publishers, Inc.
17385 Conant, Detroit, Michigan 48212



89 88 87 86 85 84 83 8 7 6 5 4 3 2 1

Library of Congress Catalog No. 83-060230
ISBN 0-910853-00-2

Manufactured in United States of America

Author David M. Osborne is an instructor at ASEA Robot Division and has served in engineering positions in many leading corporations. His involvement with robots and automated machinery began in 1968 with an attempt to develop a beer retrieval robot. Mr. Osborne was graduated from Wayne State University with a B.A. degree and has pursued post-graduate work at Wayne and The University of Michigan. He and wife Nancy have a son, Jason David.

To George Wallace Osborne—banjo player! Also, many thanks to Anita Dallaire for her help in typing and Nancy and Jason for being my cheering section.

## CONTENTS

## ILLUSTRATIONS

**(Figures with * by Randy Komraus)**

# INTRODUCTION

The popular American conception of a robot is that of an artificial man. A robot is imagined to be an anthropomorphic montage of bicycle parts and computer bits. Science fiction writers and moviemakers perpetuate an image of the robot as a creature of sublime benevolence or menacing evil. Robot characters with human traits become the subject of fascinating fiction.

However, in this story telling, the technology of robots and, for the most part, any indication of their practical use, are left conveniently untouched. But, while the make-believe was taking place in books and on the movie screen, real robots were being built and used to manufacture goods. Robot technology began in the United States and, without much fanfare, was put to use.

Gradually, however, an economic series of events was to begin a subtle change in the perception of robots in America. The United States as well as most western nations began to experience hard times. The country and the world watched as one of our largest corporations, Chrysler, teetered on the verge of bankruptcy. If this giant could not survive, was there safety for any of us? There was talk of the economic miracle in Japan. Japan was held up as a model for American economy. How was it that Japan could produce goods in such great quantity and with such quality as compared to their American counterparts? Delegations traveled to Japan and to Europe, and elsewhere if productivity seemed high. The new watchword in the media was *productivity*. We wondered constantly why the United States was losing it. The economic attitude was, we had to do something.

Some changes were made in American manufacturing. *Quality circles* were developed in an imitation of the Japanese practice of placing responsibility for quality at the actual point of manufacturing. And, as attention was focused on Japan, the striking number of robots in use there became evident. Equipment, much of which was developed in the United States, was now seen by the general public as a new development of the Japanese.

During the midseventies, there were robotic conferences and exhibitions open to the public. These shows usually were held in conference rooms of fair-sized hotels and were not overly crowded. A typical show was represented by perhaps 20 manufacturers, and the interest generated for their equipment came from only a few companies. But, by 1982 the public showed decided interest in robots. The halls used for displays, although vastly larger than in previous years, were now too small.

A different kind of people attended the shows. They were small shop owners from Tennessee who sought information about these new robots. They were investment company representatives asking questions. They were students looking for career directions. They were housewives who wondered if there were robots to be used in their kitchens. And, they were factory workers, in great number, sizing up the competition and feeling wrath or amusement at the sight of robot arms.

At one particular robot show, the crowds were so great that the fire marshall had to limit the number of people entering the hall. For the first time, the public remained at the robot booths for long periods, asking questions, watching over and over again the same demonstration.

New companies had entered the market. Major appliance manufacturers, large computer companies, and aerospace companies, already successful in other markets, chose to unveil robotic equipment. The word was out: "The robots are here."

This writer worked in the ASEA company booth during the 1982 robot show in Detroit, Michigan; and, unlike previous exhibitions, there was a feeling of awe in the air. Most of the record 25,000 people who attended had never seen a robot before. It could be seen in their faces and it was obvious by their comments that they were struck by the high level our technology had reached in this field. The comment heard over and over again was, "They are real!"

When a voice from a booth could be heard responding to questions in a knowledgeable fashion, a crowd soon gathered. The robot company salesperson was an island of information in a sea of questions. If a particular answer held the ring of economics, investment analysts soon sprung up wanting to know the potential market, which companies would do well, and whether there was still time to invest in robots.

This was no auto show or boat show with people coming in to see what was new. These were people seeking information about a new age. The exhibits were not something to dream about having in the garage. They were the new conscripts in an economic war, steel-collar workers for the factory of the future. As the people said, "The robots are real."

# CHAPTER 1

## WHAT IS A ROBOT?

It would seem that since so many people speak about robots and that so many manufacturers are producing them, there should be an agreement about what robots actually are. This is not the case. One manufacturer may say that the goods it produces are inherently different from the goods produced by another manufacturer, when in fact both have products called robots. What the buyer considers a robot is, to the manufacturer, a multitude of separate markets and products.

When statistics are compiled on the use of robots, the numbers take on a phantomlike mystique. Comparisons are made between the number of robots in use in America and Japan. When robots are counted in an American manufacturing plant, only those pieces of machinery called robots *when purchased* are included. In a Japanese facility, *any piece of machinery which performs a robotic function* often is considered a robot. Articles about productivity have stated the Japanese are using anywhere from 10,000 to 50,000 robots. American companies might well think, "How do they have so much automation in comparison to us?" But at a Japanese facility, instead of vast hordes of robots, very simple mechanisms are seen which, when used in American and European manufacturing facilities, are not considered robots. The number in use is large, but not on the scale reported. A more realistic though approximate number is shown in Figures 1.1 and 1.2.

What then should be considered a robot? With the new interest in robots, it became fashionable for a manufacturer to take existing components and machinery and label them as robotic. What previously had been considered only automation was now called a robot (Figure 1.3). Equipment that had worked only manually before was now readily fitted with microprocessor technology so that it would gain the automatic semblance of a robot. It did not seem important that there be a common definition for robots. After all, robot purchasers could determine for themselves if a particular piece of equipment would

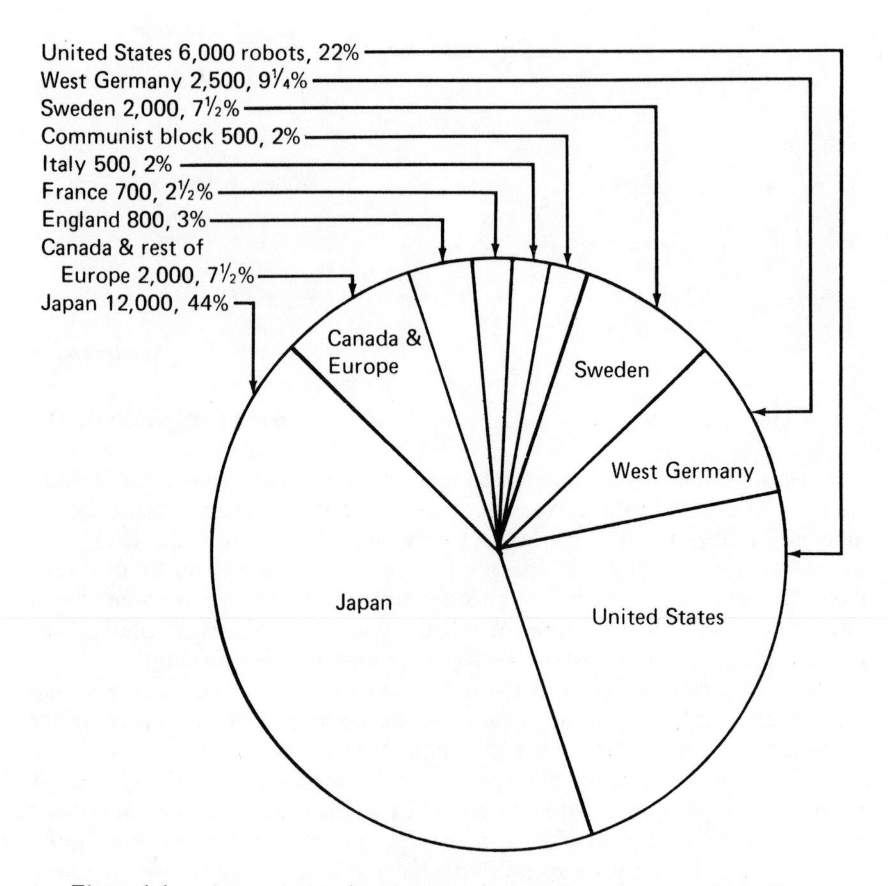


**Figure 1.1.** Approximate distribution of robots by country, end of 1982.

satisfy their needs. Whether the equipment was or was not called a robot should not influence their buying decision.

But as interest in robotics surged, when there developed an urgency to educate American industry to the potential of robots, it became important to define the basic robot.

Major trade organizations and professional societies involved in robotics published their own definitions of a robot. With legal exactness, they determined what would and would not be included in the definition. Nevertheless, the general public kept its own idea of a robot. After viewing one in action, a shop worker was asked for a definition. He responded, "It's a machine with arms and hands, and it does all your work for you."

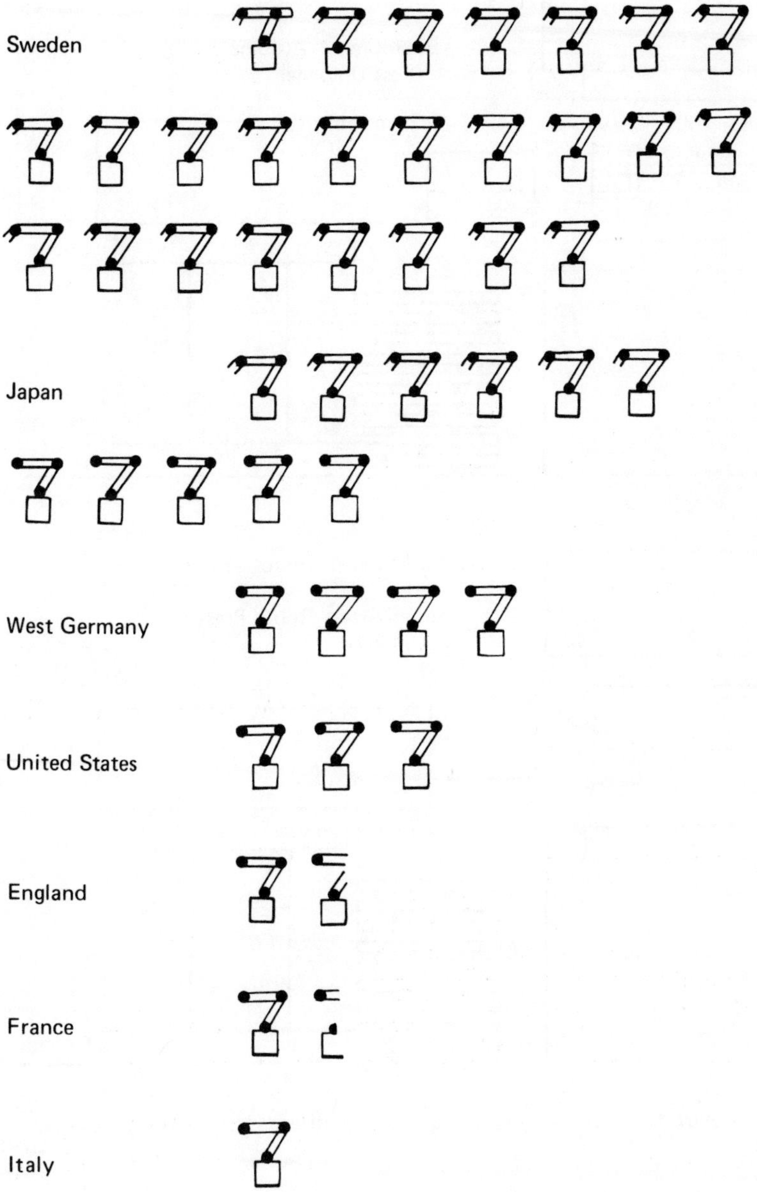

Sweden

Japan

West Germany

United States

England

France

Italy

Figure 1.2. Robots per 100,000 population, by country.

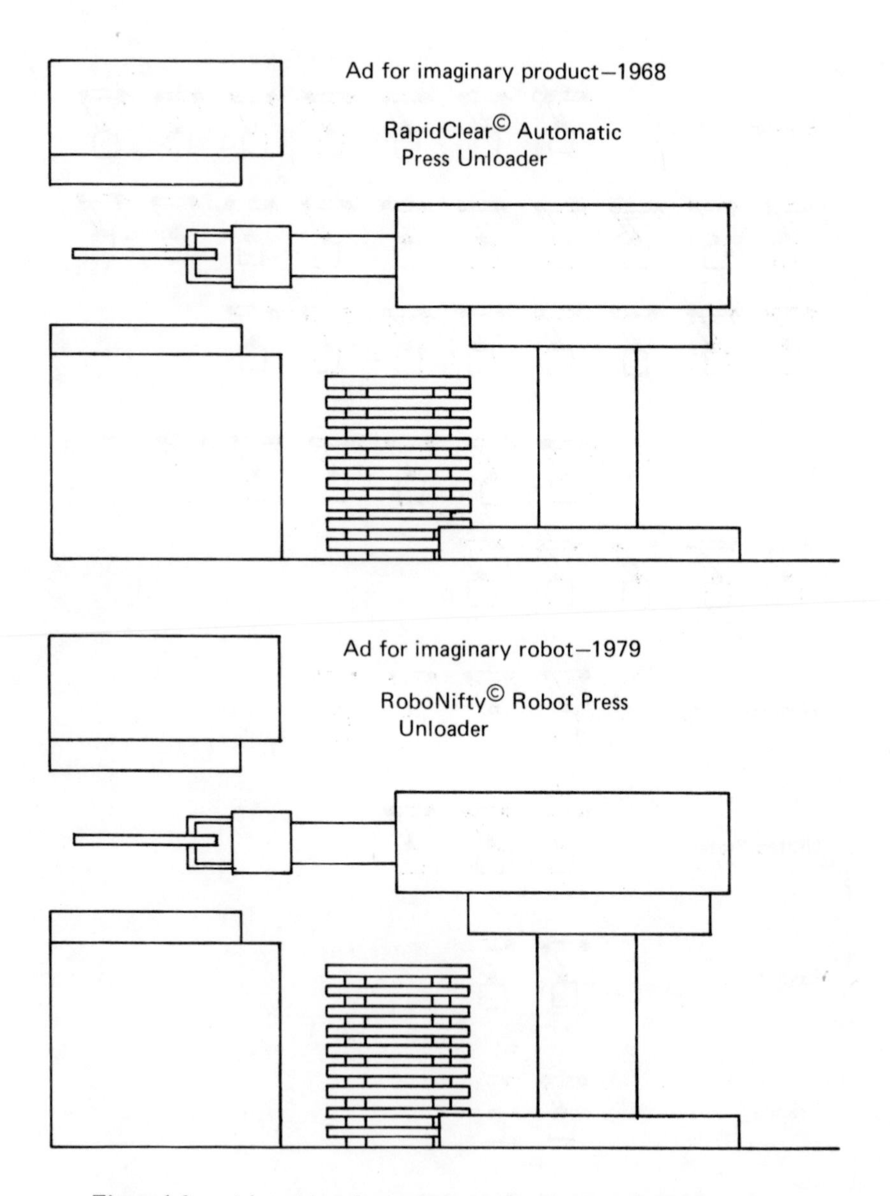


**Figure 1.3.** Advertising change to capitalize on the interest in robots.

# CHAPTER 2

## ROBOT: A SIMPLE DEFINITION

Although there might be other definitions in some ways more accurate, a simple definition of a robot is: "A machine which automatically moves items, not part of itself, to positions chosen by the robot programmer and which gives the programmer a variety of positions to choose from."

To illustrate that this definition sufficiently limits the topic to the modern conception of the robot, consider some examples that do *not* fit this definition.

An automobile is not a robot. It is a machine. It moves items not part of itself from one place to another, and the driver acts as a programmer in deciding where the machine will move these items. But an automobile does not operate automatically. The driver must be in constant command of the machine. If for even a moment the driver releases the wheel, the car will go in some nonchosen direction.

Likewise, a crane used in construction is not a robot. It can pick up and move parts, move them to a variety of positions chosen by the programmer, and do some operations without the attention of the operator. If the operator is looking away, a load being moved by the crane may be allowed to drop, the operator knowing it will fall to the ground. But this is not automatic operation.

A washing machine is sometimes likened to a robot. Here, we do have an automatic function. That is, control knobs are set to proper positions. The door is closed, and clothing inside is automatically washed. But, the machine does not move parts to positions selected by the programmer. Certainly, the washer agitator revolves or oscillates, but we do not choose the exact position to which the clothes will go. The control portion of an automatic washer is similar to that used by some robots, but the washer itself does not fit the robot definition.

Other machinery not considered to be robotic is *hard automation*. Hard automation refers to a type of machinery that is dedicated to a particular task. If a company is to manufacture Whatsits, and manufacture them in a large quantity, it is a sound investment for the company to purchase a machine specifically designed to manufacture the Whatsit. Furthermore, this might be a left-hand only Whatsit machine, the right-hand Whatsit needing a separate machine for its manufacture.

But what happens if, after a few months of production, the Whatsit market changes, and a left-hand *reversible* Whatsit is needed? The machine designed specifically to make the regular left-hand Whatsit is now obsolete. It may or may not be capable of being retrofitted to produce the new type equipment. Even if the retrofitting is possible, it is usually expensive to take the machine apart, to manufacture some new components, and to reinstall the machine for manufacturing left-hand reversible Whatsits.

Robots are sometimes called *flexible automation*. That is, if the work function or positions to which the robot will go are changed, the robot need not be remanufactured. The difference between a robot and hard automation is in the work needed for changes. Although a machine might have "arms and hands" that will physically cause it to look like the popular conception of a robot, if the unit's programming is not changeable, if it is not possible for the unit to go to a variety of positions chosen by the operator, then this is not a robot.

Another machine that does not fit the definition of a robot is a "mall robot," as seen in Figure 2.1. This machine can be found in shopping malls throughout the civilized world. Usually, there are wheels to carry the "robot" about, a body with flashing lights, several arms thrashing about, and a head with representations of eyes and mouth. A mall robot is not automated machinery. Somewhere on the periphery of the crowd, there is an operator controlling the motions of the machine. Even though many of these motions and some prerecorded voice are automatically directed by the operator, the machine will not function in any desirable way without the human operator. If this creature were to enter an industrial work place, it would have little ability to perform any useful task. As it needs the constant attention of an operator, little is gained by a manufacturer in the use of such a robot.

A simple example of a machine that fits the description of a robot can be found in the average home: an ordinary reciprocating lawn sprinkler (Figure 2.2). First, the sprinkler "moves parts." The parts happen to be droplets of water, but they are nonetheless parts moved by the machine.

Second, the sprinkler moves these parts "to positions chosen by the robot operator." In the case of the washing machine, the positions reached by the clothing were not exactly selected by the programmer, and thus the washer was not considered a robot. However, the positions

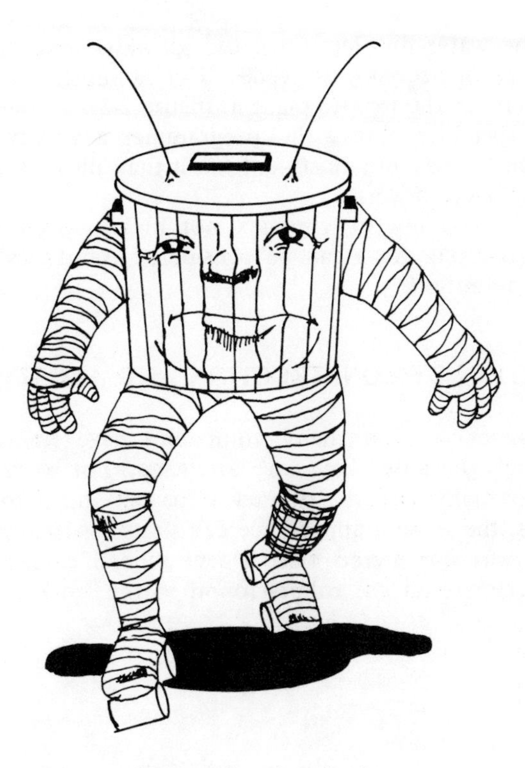

**Figure 2.1.** A mall robot.

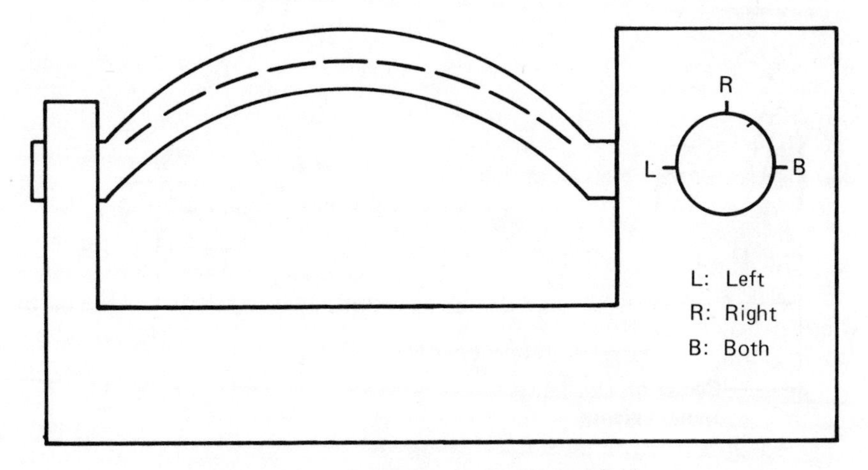


**Figure 2.2.** A simple robot: a lawn sprinkler.

reached by the water droplets from the sprinkler may be adjusted to whatever degree of accuracy is needed. For example, the droplets may be adjusted to reach the tomato plant in Figure 2.3.

Third, the sprinkler "gives the programmer a variety of positions to chose from." It has an adjustment knob that allows the water to be directed to various positions.

So, a robot need not be complex nor highly sophisticated. Many day-to-day household items can be considered robots, as long as they fit the simple definition.

## CLASSIFICATION BY CONSTRUCTION AND ABILITY

In recent years, a great many companies have entered the robot market. Partially because they have attempted not to violate existing patents and partially because of new ideas developed for ideal robot configurations, the general appearance of robots has changed drastically. Formerly, it was considered that a few general categories of robot motion approximate all the robots found in the work place, and that

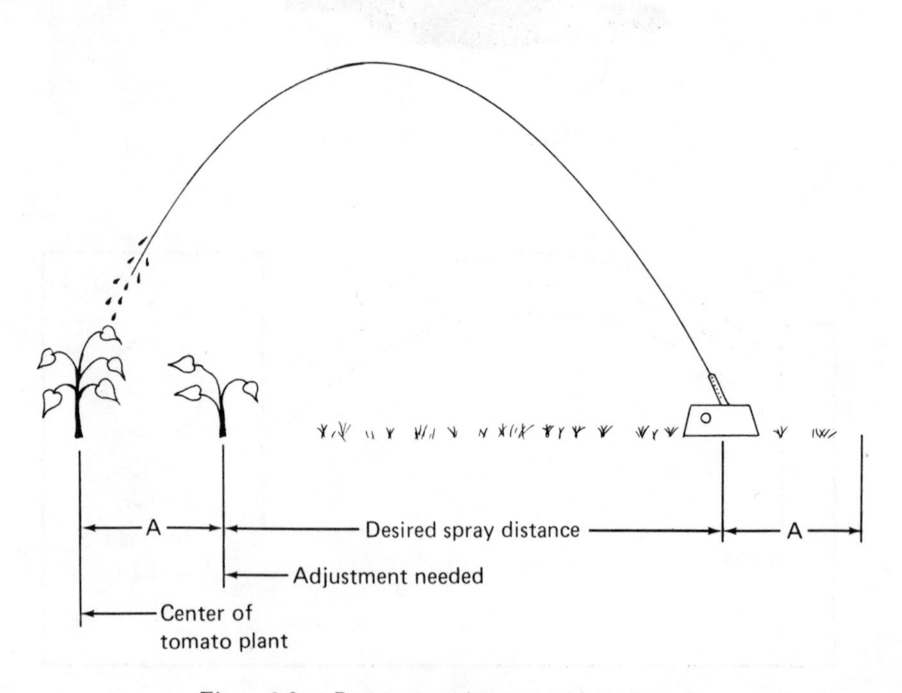


**Figure 2.3.** Positioning a lawn-sprinkler robot.

only simple or complex controls could be achieved. This is no longer true. The classification system which follows is based on a modular approach to all robot motions and activities.

## CLASSIFICATION BY MOTION

First, let's identify the basic motions a mechanical component might take. For the purpose of our classification, there are two basic motions for a robot's arm. Initially, there is the linear travel of one arm upon another (Figure 2.4). Here, one arm carries another arm along a straight path. We will refer to this motion as *linear*. The second basic motion a robot arm makes is one involving *rotation* (Figure 2.5). The arm rotates about some fixed point, usually a bearing. There are two submotions that are really extensions of these first two. Consider a robot arm where the motion itself is linear but one portion of the arm physically fits inside another portion of the arm. This type of motion is called *extensional* (Figure 2.6). Further, if there is a rotating motion but the

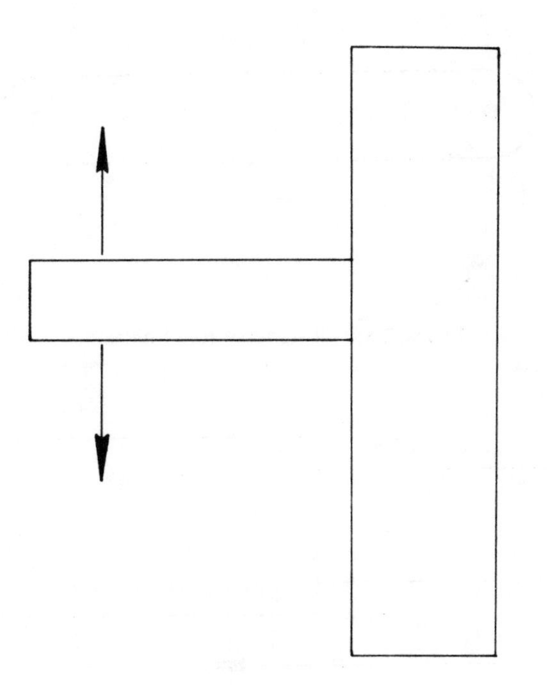

**Figure 2.4.** Linear robot motion.

center line of the rotation is also the center line of the arm itself, this is a *twisting* motion rather than a rotational motion (Figure 2.7).

These four terms then, linear, extension, rotation and twist, form the components of the LERT robot construction classification system. Figures 2.8 through 2.24 show many popular robots manufactured today and their LERT rating numbers.

In order to avoid eye confusion where several axes of the robot have the same motion, the LERT system uses a numeric superscript to indicate the number of times a particular motion is repeated. For example, RRRRRR becomes $R^6$ meaning six rotation axes.

Furthermore, the lettering always begins with the robot component mounted to a fixed surface. This is usually the base mounted to the floor but is sometimes a base mounted to the ceiling, a wall or a piece of machinery. Each axis is then listed in order as it is mounted to this first axis component. When there is some ambiguity as to the source

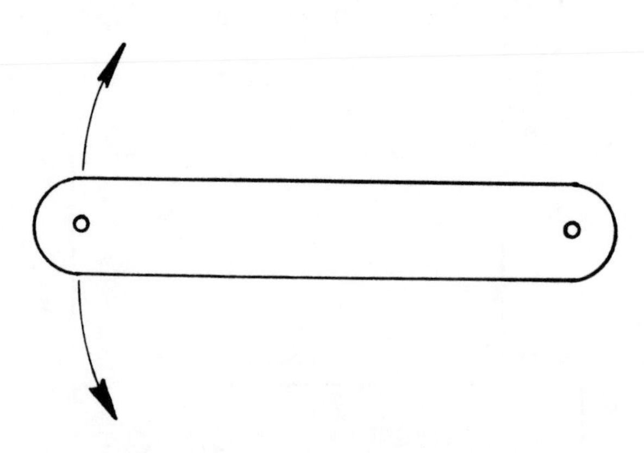

**Figure 2.5.** Rotational robot motion.

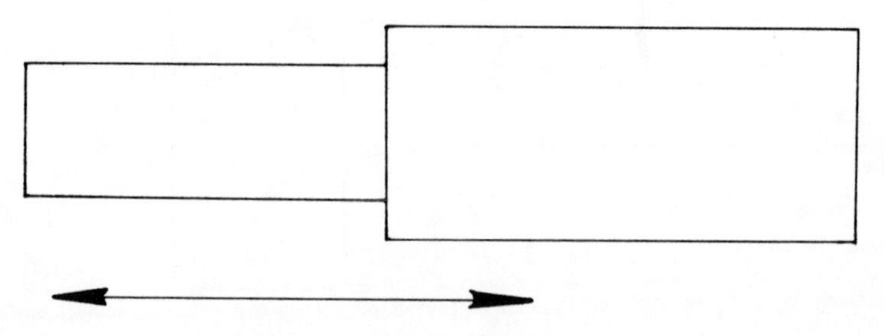

**Figure 2.6.** Extension robot motion.

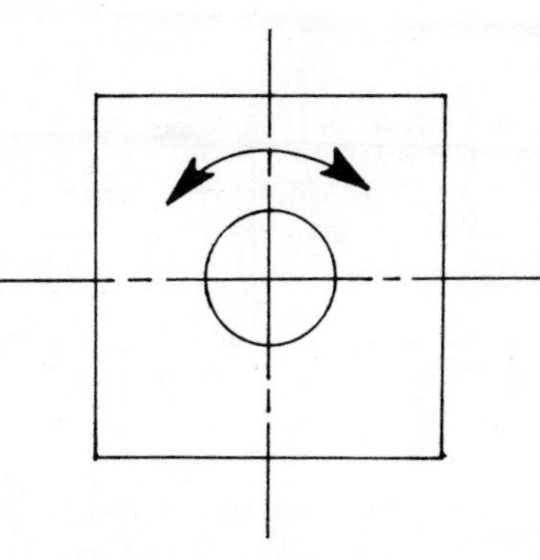

**Figure 2.7.** Twist robot motion. Note: the center line of the moving axis must be the same as the center of the motion.

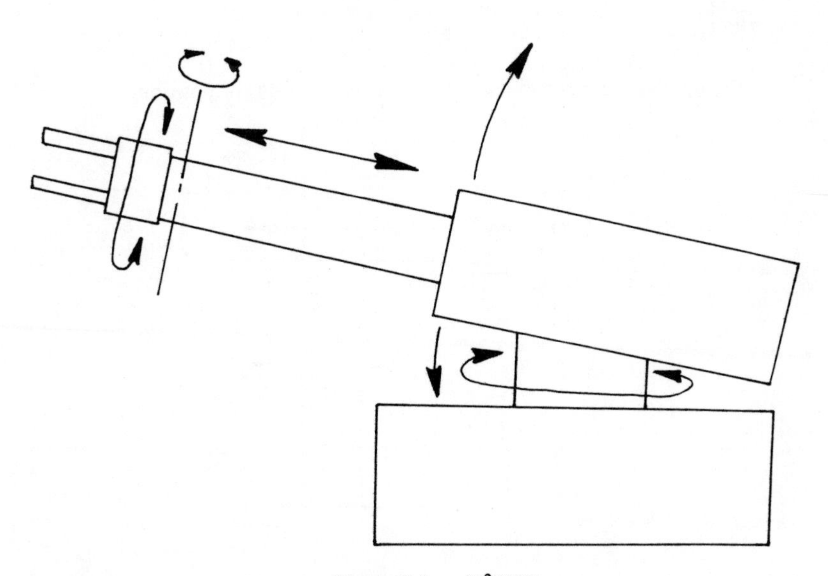

**Figure 2.8.** $R^2$ERT.

of the robot motion—whether it is servo controlled or a pick-and-place type—an indication can be made at the beginning of the letter description. If a single robot incorporates elements of both types of motion, a subscript letter S for servo or H for hard stop can indicate which of the axes is controlled by each method of motion.

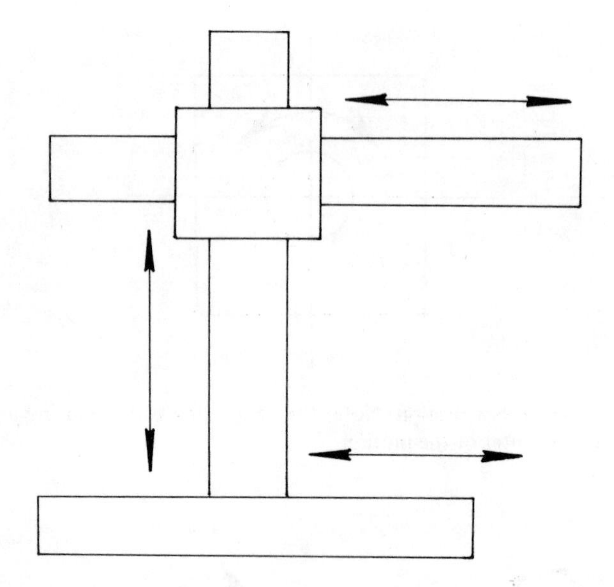

**Figure 2.9.** $L^3$. Note: base motion shown 90° out of position.

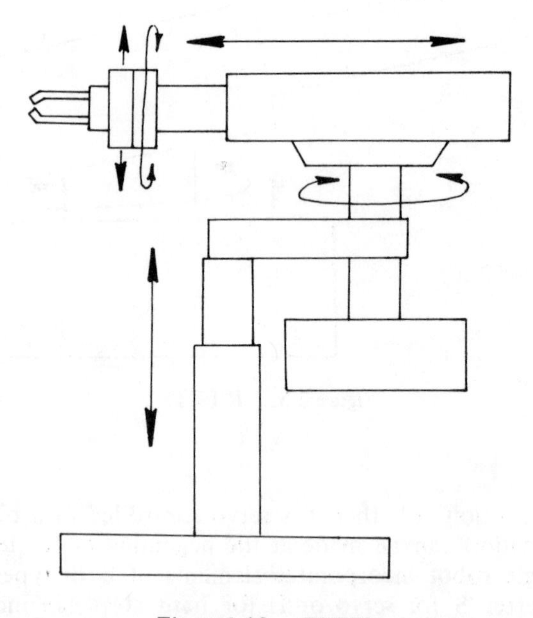

**Figure 2.10.** ERETL.

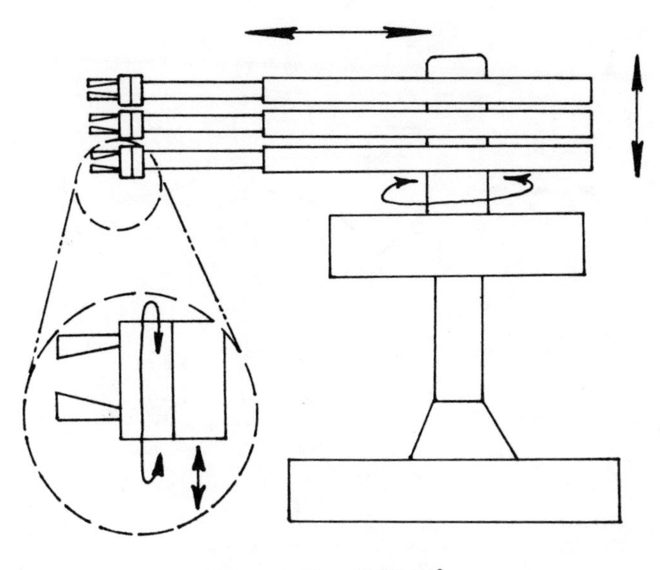

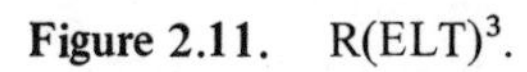

**Figure 2.11.** $R(ELT)^3$.

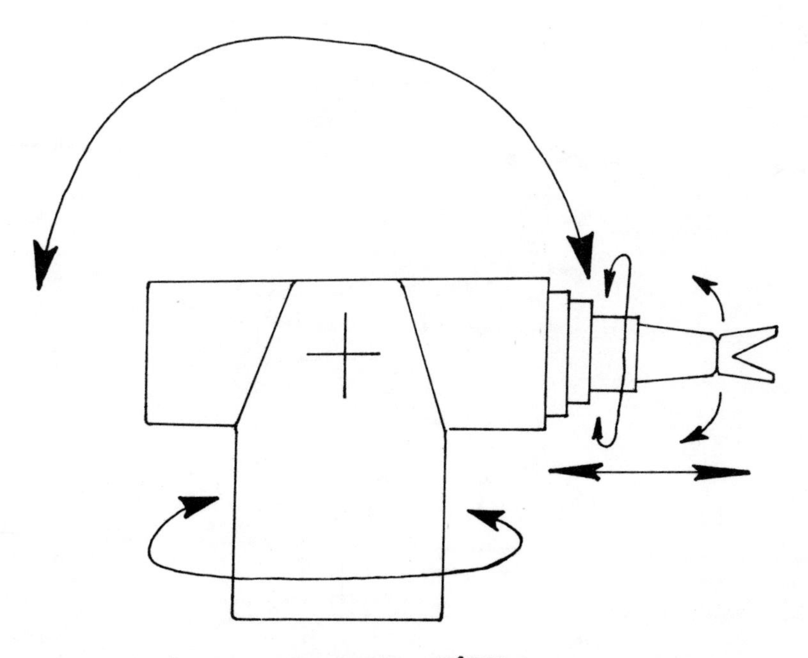

**Figure 2.12.** $R^2ETR$.

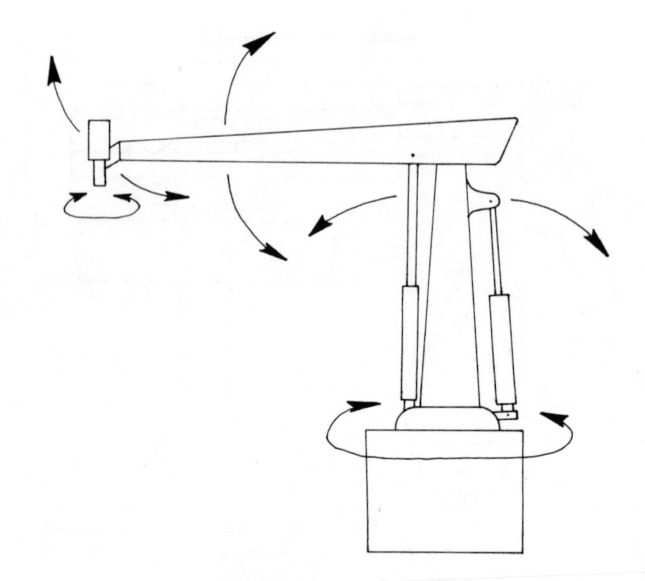

Figure 2.13. $R^4T$.

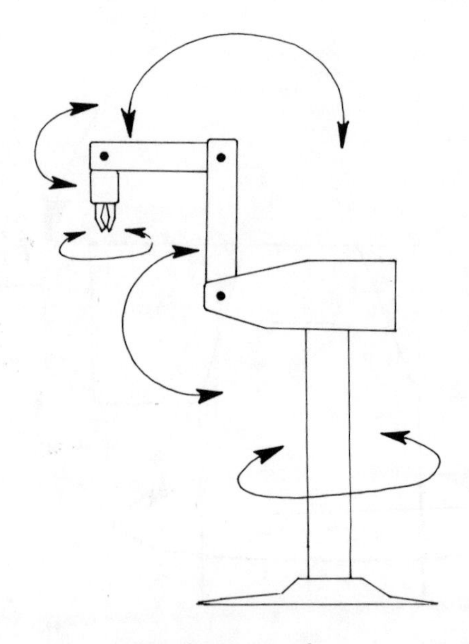

Figure 2.14. $R^4T$.

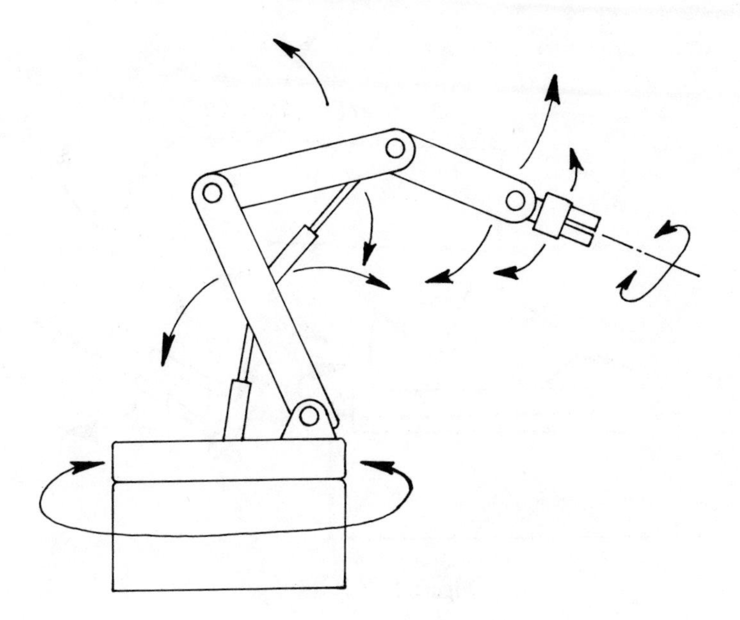

**Figure 2.15.** $R^5T$.

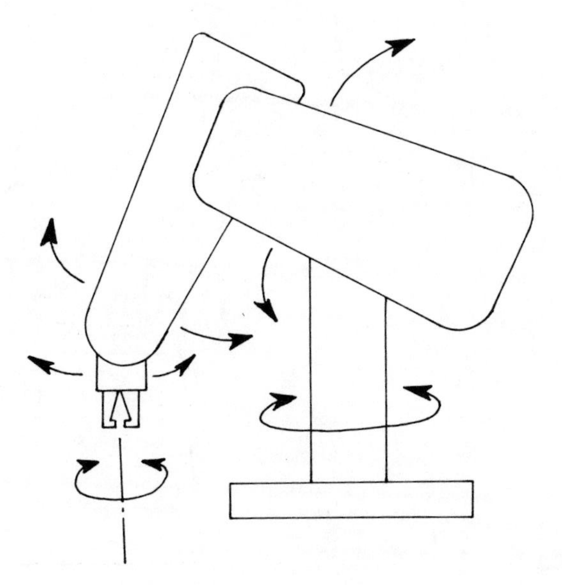

**Figure 2.16.** $R^4T$.

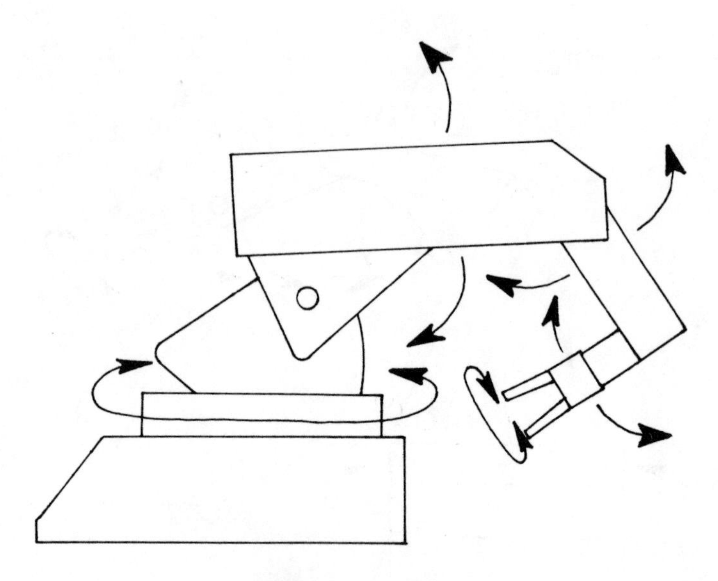

Figure 2.17. $R^4T$.

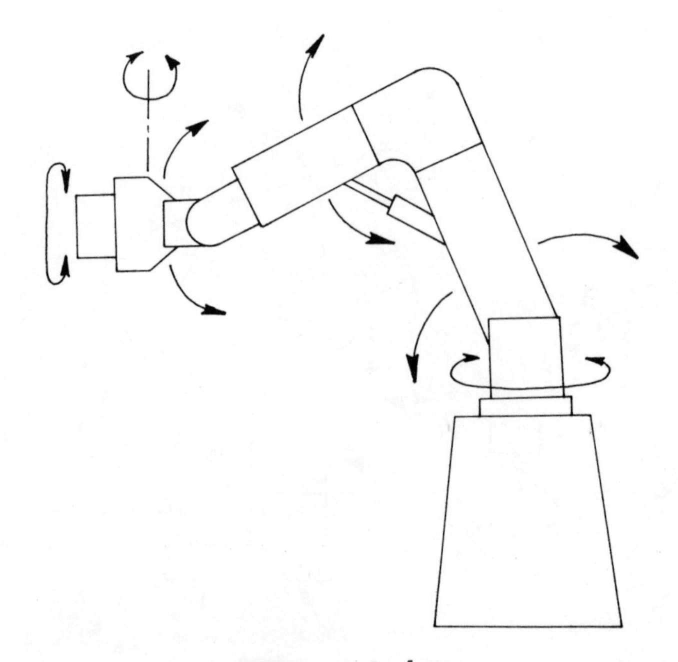

Figure 2.18. $R^5T$.

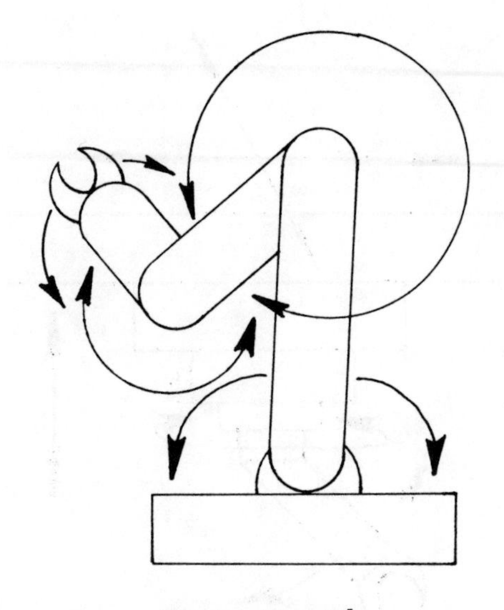

Figure 2.19. $R^5$.

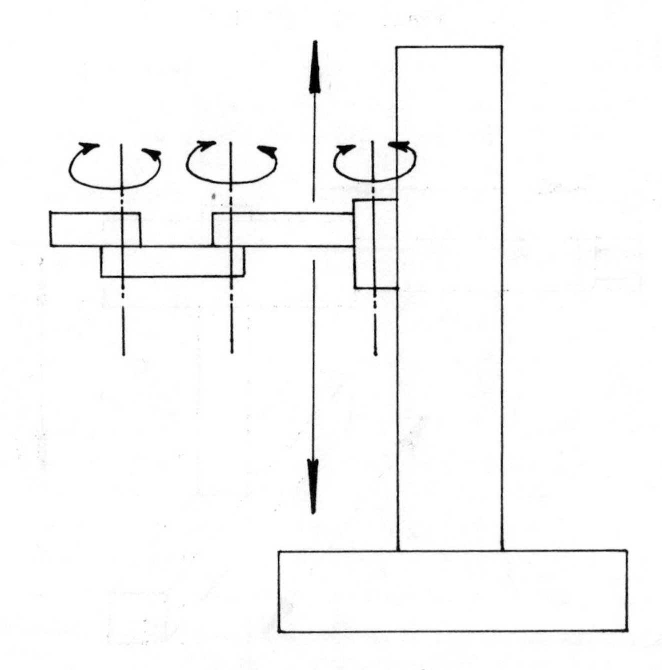

Figure 2.20. $LR^3$.

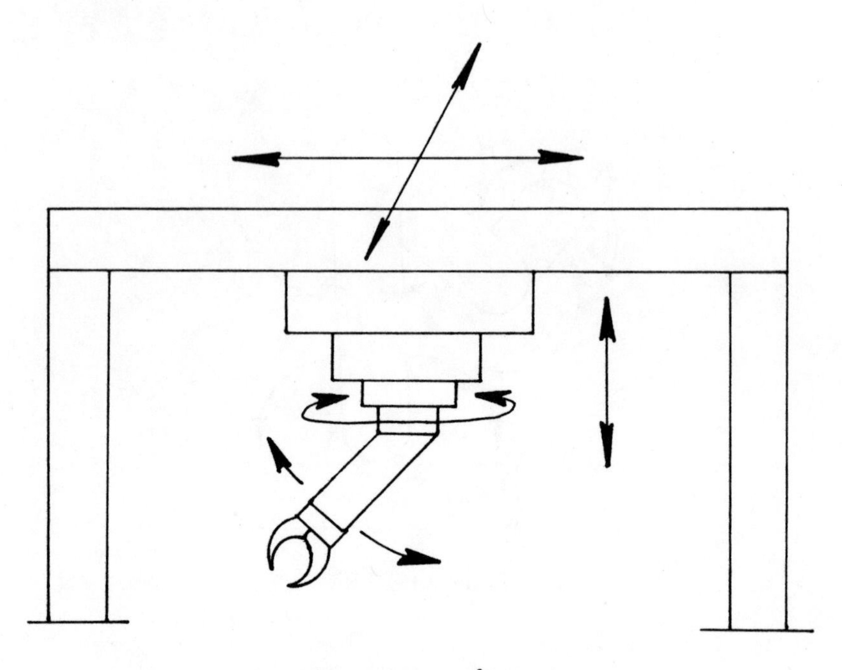

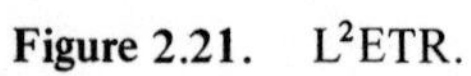

Figure 2.21. $L^2ETR$.

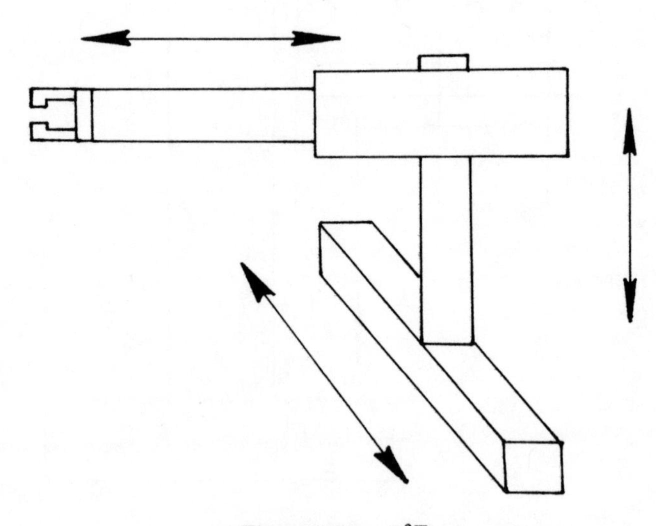

Figure 2.22. $L^2E$.

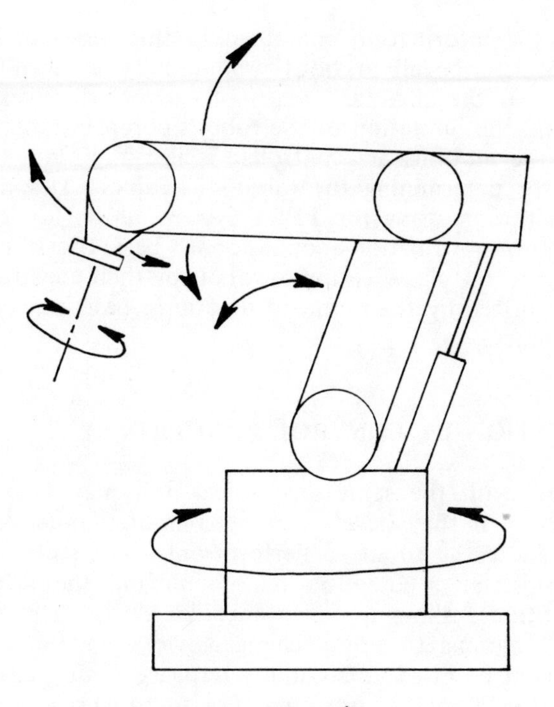

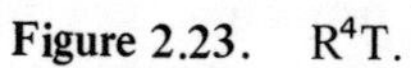

Figure 2.23. $R^4T$.

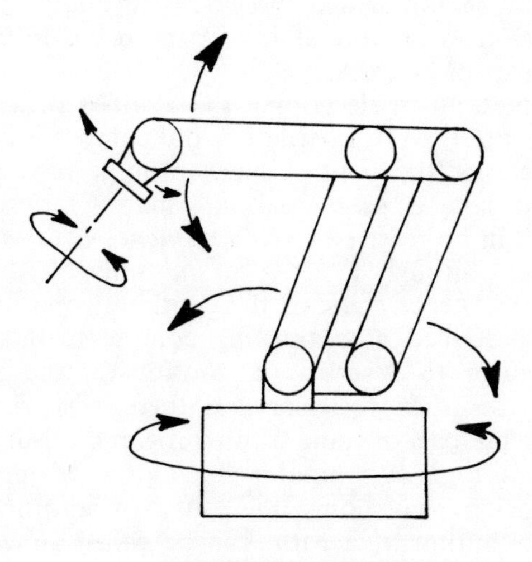

Figure 2.24. $R^4T$.

Certainly, the information contained in this series of letters is not adequate to describe all motions produced by a robot. Information is needed about the angle and reach of each of the axes, the robot's work area and the limitation of the robot's complex motions. The ease with which the motions are controlled is also important to a potential robot user for determining the robot's usefulness. Despite the limitations, a description using the LERT system allows for less ambiguity in the transfer of information about a robot's basic construction.

In addition to the classification of robots by their construction, robots may be categorized by their control functions, basic power supply, and physical accuracy.

## CLASSIFICATION BY CONTROL FUNCTION

Two robots with the same mechanical unit may be very different in their abilities if they have different computer units. A simple computer may allow the robot to perform only rudimentary tasks, while a more sophisticated computer may command the robot into very complex motions, allowing for humanlike decision making and the altering of programs to suit a changing work environment. A robot computer might be rated on its ability to process information or control memory for the storage of programs. For those who are knowledgeable in the technology of microprocessors, it might be enough to specify the exact computer hardware used in the construction of the robot and the memory made available to the computer. Information on technical details can be misleading, however, depending on how well the equipment is designed and the ability of the robot to use the computer efficiently as a part of its work.

A workable method of describing the abilities of a robot computer is by indicating the type of instruction the computer is able to receive. If the computer (or the control mechanism if it is not a computer) is only able to receive sequential information altering the position of the arms, it can be referred to as a *sequence-controlled* robot. When the robot is led through a series of motions by the operator, and the controller only keeps track of the continuous positions of the arms, this might be described as a *tracking* computer. If a control system allows the operator to describe the motions of the robot by jogging the arms to the correct positions and then registering them for use later as an alterable part of some future program, this might be classified as the *registration* type. A computer that allows the operator to describe the robot's motions with only the entry of information, such as a mathematical definition of a path, can be called an *information* type. For examples, see Figures 2.25 through 2.28.

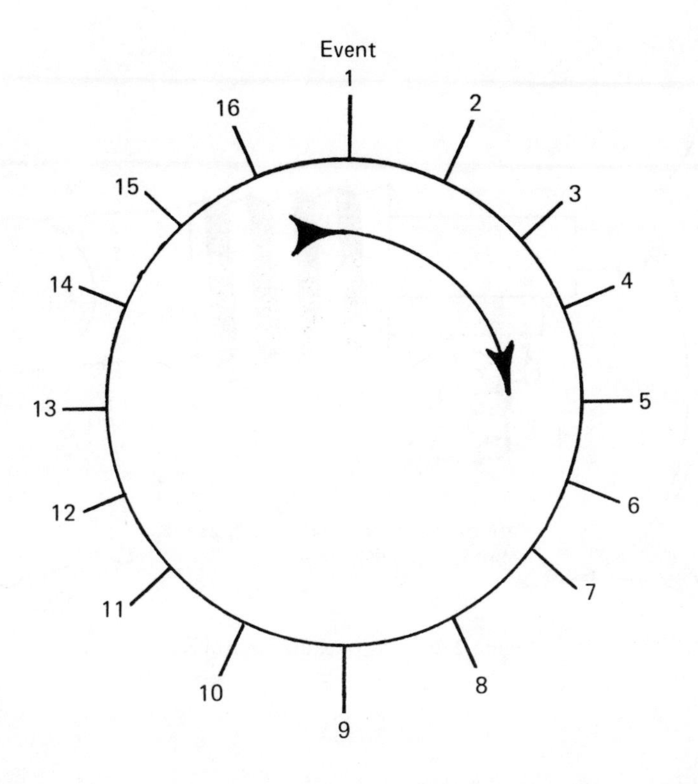


**Figure 2.25.** A sequence control only activates a set series of events.

The initials of STRI, for sequence, track, registration and information, can be used to describe the robot's command receiving ability similar to the LERT description of the physical motions. If we add to this the ability of the control system to keep in storage the individual movements that comprise the program, we have a rather simple way to rate the robot's computer abilities.

Two examples of this rating are: A robot programmed with instructions consisting of the Cartesian coordinates of the positions it is to reach, and which will hold 3,000 positions in memory, being described as I-3000, and a robot that is sequence controlled by a rotating drum with 24 positions being described as S-24.

One of the more useful distinctions to make of robots is their ability to follow exactly a determined course—the route of the robot being controlled for speed and accuracy. This is a feature of almost all

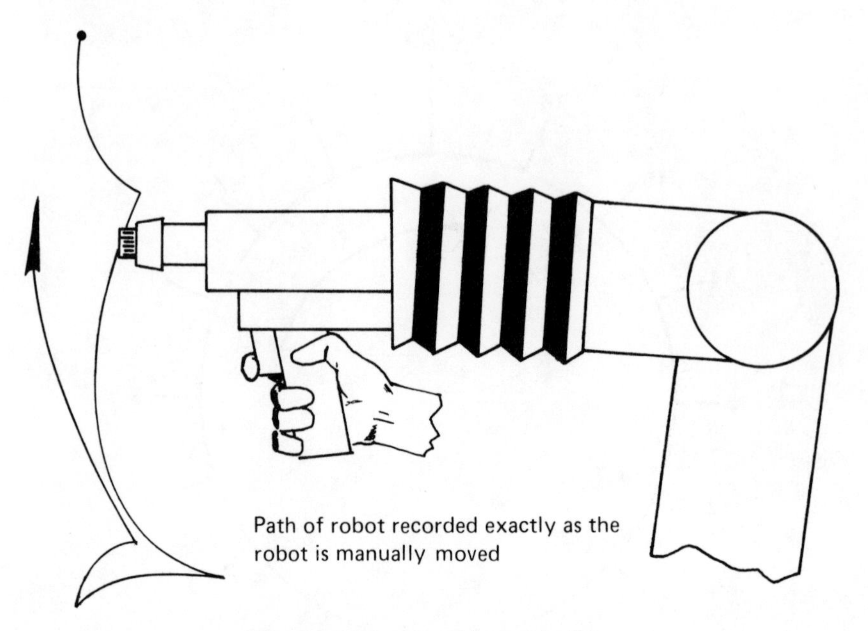


**Figure 2.26.** A tracking controller.

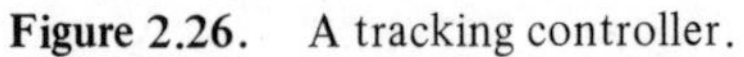

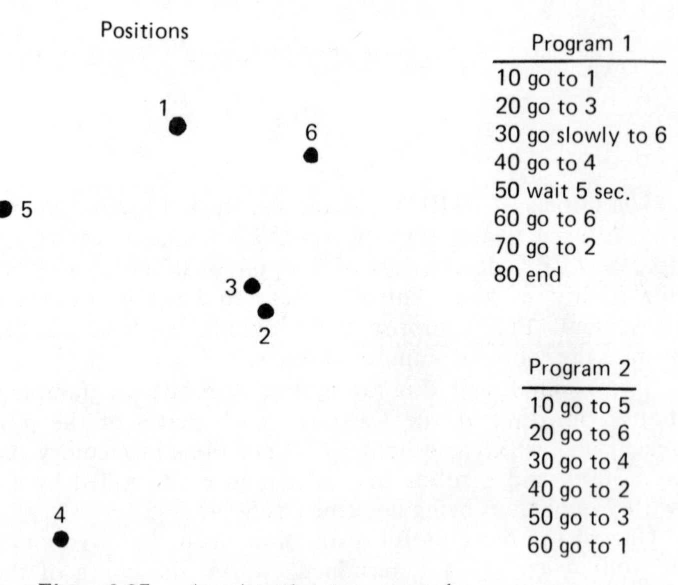


**Figure 2.27.** A registration-type control.

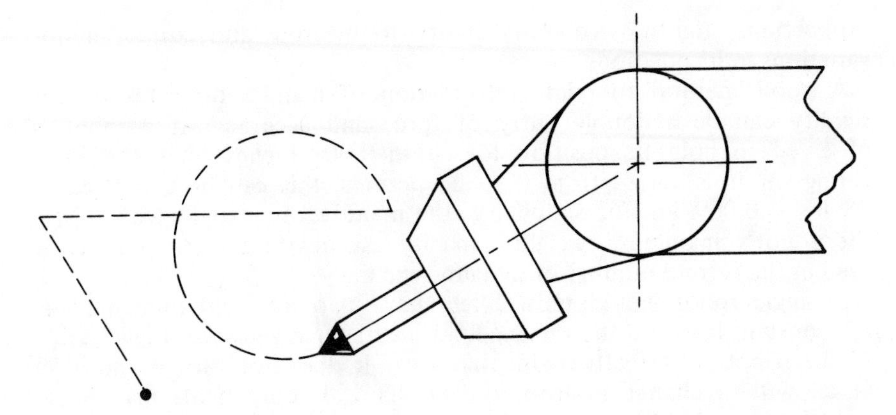

**Figure 2.28.** Informational-type controllers can receive information and build a complete program from it.

registration- and information-type robots and almost no sequence- or tracking-type robots. It is not necessary to make a separate classification for this feature.

## CLASSIFICATION BY POWER SUPPLY

There are three basic power supplies used in modern robots. There are robots powered by air, called *pneumatic*, by oil, called *hydraulic*, and by *electricity*. It is important to include the basic power supply information when describing the robot to someone unfamiliar with its basic functions.

## CLASSIFICATION BY ACCURACY

A robot's ability to reach a particular point in space exactly and its ability to hold its position against a varying load can be two of the most significant aspects of its potential. A robot's accuracy and rigidity must certainly be included in any rating of its ability. It is rather simple to indicate the ability of a robot to reach a point in space repeatedly. A number such as 0.005 inches can indicate the maximum error expected when the robot tries to reach some point. It is not so easy to rate the robot's ability to hold that position if the forces acting upon it change. A robot's rigidity is a factor of both its mechanical proportions, tightness, and its ability to measure and correct minor variations in its positions.

A good method to relate information of manufacturer testing for rigidity can be a double entry of force and accuracy. If a robot is observed to hold its position to within 0.009 inches when the force acting on it is varied from 0 to 25 pounds, this can be expressed as 25 lbs = 0.009 in. For simplicity, the manufacturer might wish to use the robot's maximum weight capacity as the variable weight factor used in the second number of dynamic accuracy.

A single robot would most likely have two different numbers, for the constant load and the varying load accuracy, as shown in Figure 2.29. If the robot is perfectly rigid, that is, if it does not vary its position at all with a change in load, the mechanical components may be so tight as to preclude any motion at all.

If a manufacturer were to describe his equipment as a servo controlled $TR^3T$, I-6000, electric, 0.008, 13 lbs = 0.016 in. robot, we would know that the robot had these features: five servo-controlled axes with the base having a twisting motion, the next three axes each rotating and the final axis twisting; a computer that allows the programming information to be entered directly as a set of numerical values; the ability to hold 6,000 positions in its memory at one time; power supplied by electric motors; a constant load accuracy of 0.008 inches and a variable load accuracy of 0.016 inches at a variable load of 13 pounds.

With this information, we would know if the basic parameters of the machine fit the task we have in mind. After this initial review of the robot's potential, we can, if necessary, ask the manufacturer for more details.

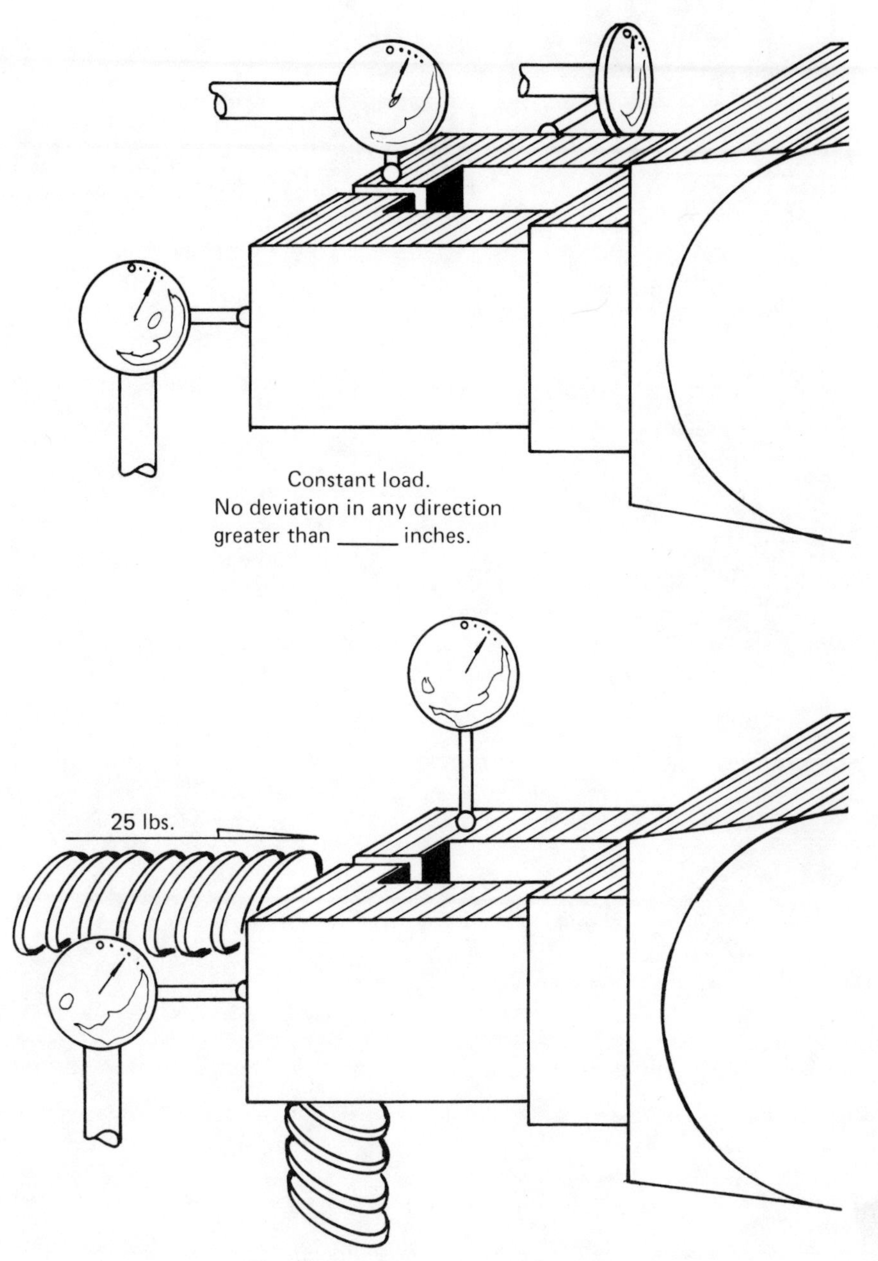


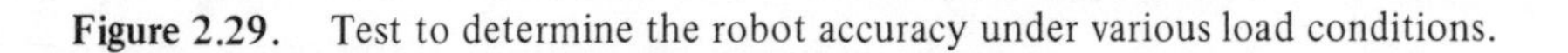
Test for deviation under variable load.

**Figure 2.29.** Test to determine the robot accuracy under various load conditions.

# CHAPTER 3

# ROBOTS CAN BE VERY SIMPLE

Robots seem mysterious to a layman but are in reality very simple. A few common elements, correctly combined, form the basis for all robot motions. The simplest type of robot, also the first historically, is called a *pick-and-place* robot and is a very simple arrangement of mechanical components.

The name pick-and-place comes from the usual task assigned to such robots. They are to grasp an object at a particular position, pick it up and place it somewhere else. The most common elements used to construct a pick-and-place robot are *air cylinders*, *hydraulic cylinders*, *solenoids*, *electric motors* and *mechanical linkages*.

## MOTION PRODUCING COMPONENTS

### Air Cylinders

Air cylinders are used for a great many purposes in industry. They are composed of a piston sliding inside a cylindrical chamber. A steam engine uses a cylinder and piston to turn its wheels. An automotive engine uses a piston arrangement to produce power. The actual piston component is simple. As shown in Figure 3.1, it is a rod joined to a circular disk. The disk has seals around its outer edge that allow it to slide back and forth in the cylinder and not let air pass over the seals.

By putting air in one end of the cylinder, we cause the piston to move toward the other end of the cylinder. In this way, the air cylinder is capable of producing two motions. One, a motion towards the rod end, and two, a motion away from the rod end. It is usually not feasible to use an air cylinder by itself to produce any other motion, for example, to move only half way toward one end. Air is not a reliable source for positioning. Unless some other component stops the piston

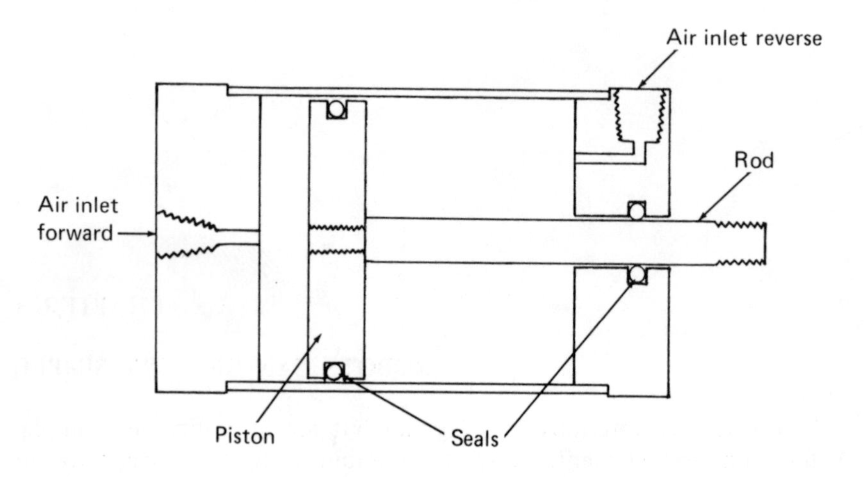


**Figure 3.1.** A pneumatic cylinder.

at the correct position, the compressibility of air causes the piston to move when any force is applied to the robot.

### Hydraulic Cylinders

Hydraulic cylinders are physically very similar to air cylinders. The differences are basically in the materials used in construction. Hydraulic pressures are in great excess of normal air pressures. A standard hydraulic pump unit will usually produce pressures in the 2,000-pound-per-square-inch (PSI) range. Most air pressures used in industry are less than 100 PSI. An air cylinder with its seals intended for 100 PSI will usually destroy itself after even a short time if it is used with hydraulic pressures.

There are some other differences in the way hydraulic cylinders work, and these are based on the viscosity of the hydraulic media. Oil will not move through as small an opening as will air. Therefore, when hydraulic oil is used, it is necessary to have the openings into the cylinder, and also the valve which controls the cylinder, larger.

To control both an air and a hydraulic cylinder, it is necessary to have a valve. A valve is used to select the cylinder motion we wish. When a valve routes either the air or hydraulic oil in one direction, it exhausts air from the other direction (Figure 3.2). A simple signal from the robot controller can shift the valve.

The speed of a pick-and-place robot can be controlled in one way

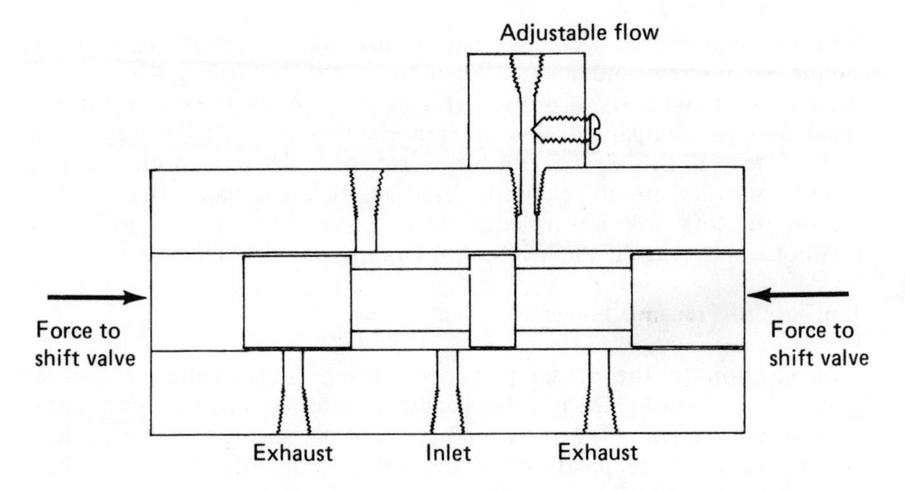


**Figure 3.2.** An adjustable flow valve.

by regulating the ability of the valve to exhaust one side of the cylinder. When this exhaust is regulated, the cylinder can proceed no faster than its ability to push air or oil through the metering device. By changing the metering device, the speed of the arm of the robot can be changed.

## Solenoids

Solenoids are special purpose electromagnets. When electrical current is put through a coil of wire, a magnetic field is established. By properly selecting an iron core for this electromagnet, we can produce a linear motion that results from the electromagnetic attraction. Solenoids, however, are limited in their ability to move a robot arm because the amount of force produced by this electromagnetic attraction is small in comparison to the amount of electricity used to produce that force. Solenoids are used in some robots for simple tasks, particularly when these tasks must be done in a very small area and involve very light loads.

## Electric Motors

A standard electric motor is sometimes used in a pick-and-place robot; but, because the motion produced by the electric motor is rotary and also because there is little control of the actual position of an electric motor, these motors are usually used only when coupled with some type of transmission element. This transmission element might be as simple as a screw and a nut. When the screw is turned, the nut will produce some linear-type motion (Figure 3.3). When the electric motor turns and its companion transmission element goes all the way to its end of travel, the electric motor is stopped. This is another reason electric motors are infrequently used in pick-and-place robots. Most electric motors will not tolerate such a stop. The wire coils within the motor may burn if the motor stops and the power is left on.

## Linkages to Transmit Power

In addition to the actual power-producing components such as air cylinders, pick-and-place robots usually have some type of power transmission component. The individual element producing the motion may be located at some position on the robot body that does not allow the individual robot arm access to its work piece. In this case, the motion must be produced in one part of the robot and actually used in another.

In a bicycle, to use an example, the actual power to turn the wheels is generated by a set of human legs (Figure 3.4). Because the pedals

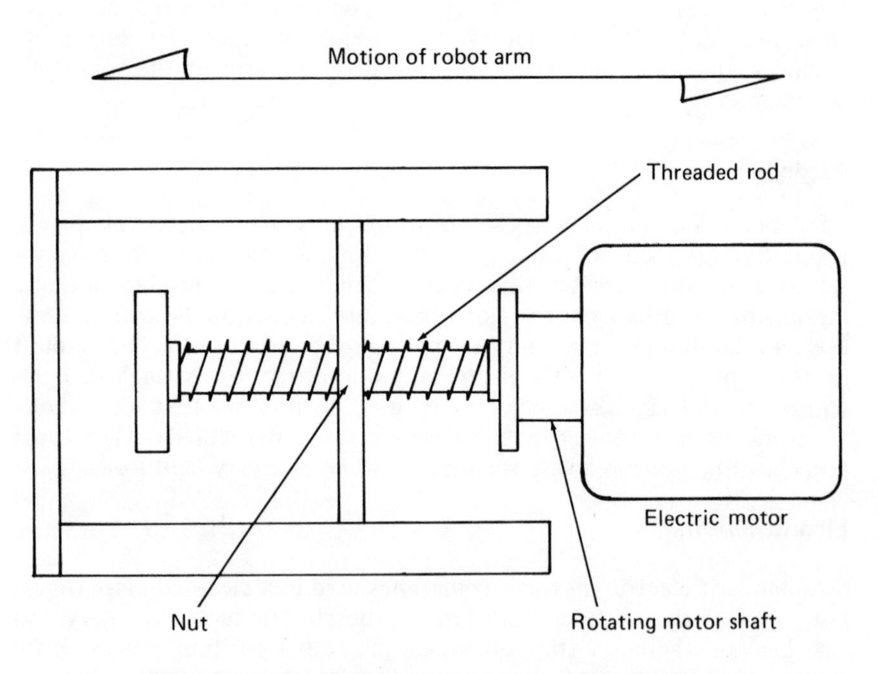


**Figure 3.3.** Simple linear motion using an electric motor.

**Figure 3.4.** A bicycle example of power location and transmission.

are located at a point convenient to the rider, specifically between the two wheels, it is necessary to transmit that energy to another part of the bicycle. The energy is needed at some point along the axis of the rear wheel; and so, to transmit the energy, a bicycle chain is used.

A bicycle chain is one of the common elements used to transmit energy in a robot. Although it is not a set of human legs producing the energy, the chain works in exactly the same fashion. If the end of a robot arm is to contain a gripping unit and this gripping unit must enter a small opening to do its work, it is often more convenient for the power device to be located farther back on the arm. Only the gripping unit with an element such as a bicycle chain will actually enter the opening (Figure 3.5).

**Rotary Pneumatic Elements**

Sometimes it is necessary to produce a rotary motion with a pneumatic element. By use of a special set of seals that will pass through a chamber in approximately a semicircular shape, it is possible to use air to produce a rotary motion. If we envision the chamber containing the moving element, curved on the top and bottom and straight along the sides such as the shape a windshield wiper makes on a rainy window, we can see the path that the moving element would follow (Figure 3.6). It is fixed and rotating about an axis just as is the wiper blade. As air

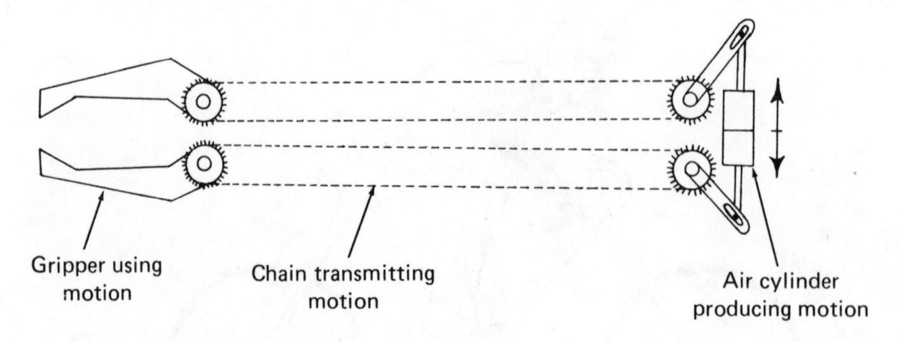


**Figure 3.5.** A robot gripper with remotely located power.

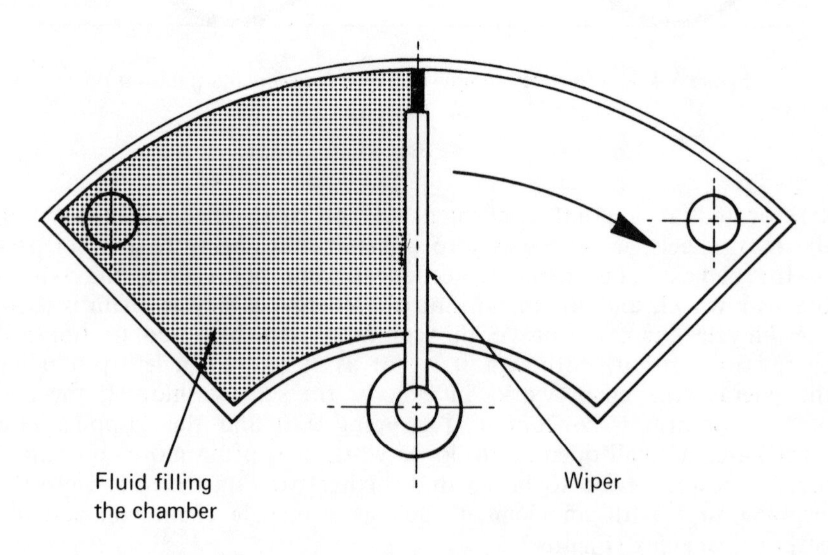


**Figure 3.6.** A wiper-type fluid motor.

is put in one side of the chamber, the buildup in pressure would cause this blade to move to the opposite side. When it has completed this partial rotation, it is necessary to apply air on the opposite side to bring it back. The wiper element and the blade element have seals on all four sides where contact is made with the chamber. It is possible to use this arrangement to move some of the robot arms in a rotary fashion.

If it is necessary for a rotary pneumatic element to complete more than one full rotation, then an air motor is necessary. An air motor is a series of wipers that travel in an eccentric chamber. We see in Figure 3.7 a chamber filled with rotating wipers. By properly positioning an inlet port and an exhaust port, we can cause these wipers to rotate continuously.

As compressed air is forced into the inlet port, the air expands against the nearest wiper blade. This pressure on the blade causes it to move toward the larger side of the cavity. As this first section of air makes its motion toward the exhaust port, a new wiper seals the portion of the chamber at the inlet port. The first wiper then passes over the exhaust port, allowing its compressed air to escape. The cycle is complete and the second wiper repeats the actions of the first. Because each wiper returns to the starting point after it has completed its task, the motor may turn in a continuous fashion similar to an electric motor.

A distinct advantage in the use of air motors, as opposed to electric motors, is the size necessary to produce a given force output. Since

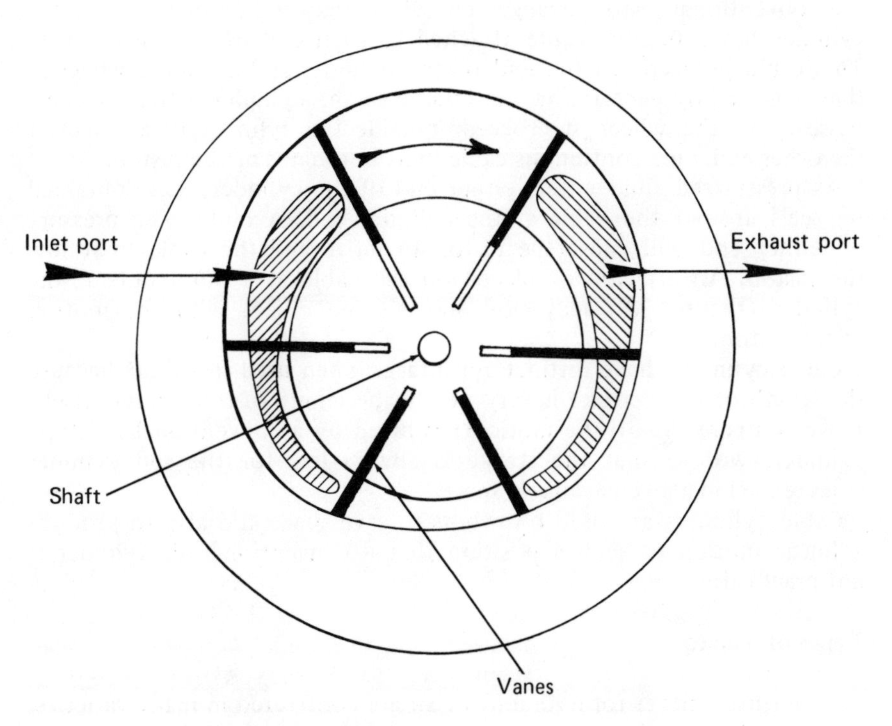


**Figure 3.7.** A full-rotational type fluid motor.

robots often are turning slowly, it is necessary either to have a rapidly turning motor with a mechanical transmission reducing its output speed or a motor that turns slowly directly connected to the robot arm. Electric motors are not well suited to turn slowly. The magnetic impulses producing the force of an electric motor must change rapidly to produce the motor's maximum output. In contrast, an air motor may turn quite slowly and have its full available force at all times. The overall size of the electric motor and its transmission is greater than its pneumatic counterpart. Another factor that speaks well for fluid motors is their ability to be stopped while still under power. If some obstruction in the path of the robot impedes its motion and the motor stalls, no harm comes to the air motor. In an electric motor, when the power is left on and the motor is stalled, electrical current in the motor windings can quickly heat up the fine wires and cause them to burn out.

### Cable Cylinders

A cable cylinder is a special type of air and hydraulic cylinder. Unlike the conventional piston arrangement with a rigid rod on one end, a cable cylinder has a flexible cable attached to each end of the piston disk. This cable proceeds to the end of the cylinder and around a wheel so that it is exactly centered in the middle of the cylinder. After the cable passes over the wheel, it proceeds outside the cylinder to a wheel at the other end. One continuous cable joins both ends of the piston.

As pressurized fluid enters either end of the cylinder, it is contained by seals around the cable at the cylinder end. A buildup in pressure on either end will cause the piston to move and the cable to follow the piston. An attachment placed on the cable at a point outside the cylinder (Figure 3.8) may now move some object with the motion of the piston.

Cable cylinders have distinct advantages when used in robots because the length of their stroke is very nearly the length of the cylinder itself. If we compare this to the motion produced by a conventional rod-type cylinder, we see that the stroke is always less for the rod cylinder (Figure 3.9) in a given available space.

Cable cylinders are used on robots to save space and also to produce a linear motion to such a position that a conventional air cylinder is not practical.

### Types of Valves

Valves used either for hydraulic or air are configured in many varieties. It is possible simply to turn the power supply on or off. This is like

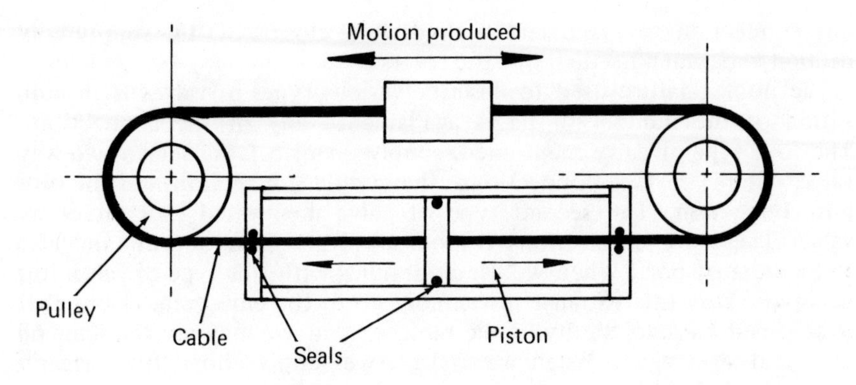


**Figure 3.8.** A cable cylinder.

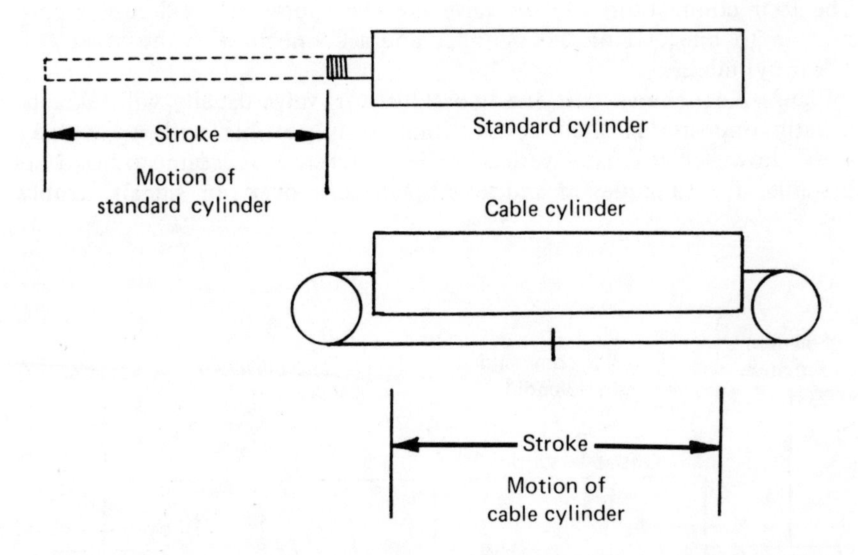


**Figure 3.9.** Cable cylinder stroke compared to conventional cylinder stroke.

a light switch, or, in terms of fluid power, a water valve. In most cases, however, it is necessary, when we stop allowing fluid to enter one end of our devices and want the robot to move in the opposite direction, to allow that fluid to escape. This is the case with an air or hydraulic cylinder. We must have a valve which serves a dual purpose: opening one connection to evacuate the fluid while closing off the supply connection and, when needed, just the reverse.

The nomenclature used to describe various types of valves is an indication of the number of paths available to any of the connections. The first type of valve mentioned, simply on or off, is called a two-way valve. There are two connections: the supply connection and the output connection. The second type of valve mentioned is a three-way valve. There are connections to the supply, to the output, and also to an exhaust port. When we use hydraulics with this type of valve, it is necessary that this exhaust not simply go to the environment but that it be piped back to the hydraulic tank, so that we may use the same oil over and over again. When we use air, we simply allow this formerly compressed air to escape into the atmosphere.

If we have an air cylinder that is alternately to move in one direction and then back to its original position, we often use a four-way valve. We may use this one valve to control all of the motions of the cylinder. The four connections of this valve are the supply, the exhaust, a connection to one end of the cylinder and a connection to the other end of the cylinder.

Figure 3.10 shows that the two states this valve usually will take are exactly opposite in their connections. It is possible with a four-way valve, however, to have various other intermediary connections. It is possible, for example, in addition to crossing over our signals, simply

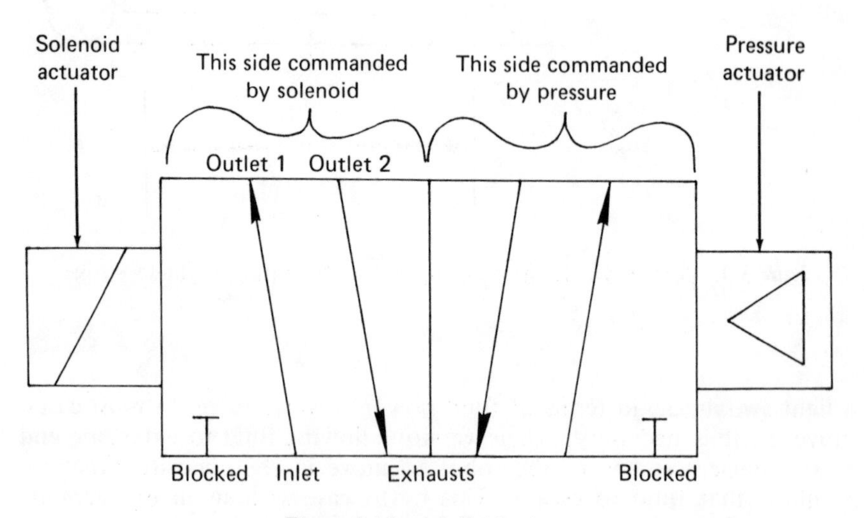


**Figure 3.10.** A fluid valve.

to trap our air or hydraulic fluid in each end of the cylinder and connect our supply to the exhaust.

In the use of hydraulics, especially, this allows the piston to be fixed in one position. Since no fluid can leave either side of the piston chamber and hydraulic fluid is incompressible, the piston itself, and hence the robot arm element it is connected to, is held rigidly in place. The connections made to the control unit, which stimulate the valve and cause it to act in one direction or the other, also can cause the valve to be in this intermediary position. If a power failure occurs, this type of valve sometimes can be used to cause the robot to maintain its position even though there is no power.

Valves that control fluid media are usually themselves controlled by an electrical signal. A solenoid is placed at one or both ends of the valve; and, when the robot controller supplies electricity to the solenoid, the valve functions. In this way, an electrical signal produced by the computer can be used to control hydraulic or pneumatic pistons. When a robot is moved manually to position by the push of a button on the controller, the human operator is indirectly causing the valve to operate. In some pneumatic robots, the control system may be assembled from pneumatic logic components. In these cases, the actual signal to the robot valves can be an air signal.

**Spring Loaded Motion**

In some instances (on a pick-and-place robot), it is necessary to have only one-half a motion produced by air pressure. If we wish to clamp an object using an air cylinder and at some later time unclamp it, it is unnecessary to use air pressure to perform both motions. A spring capable of pushing the piston to one end can be installed inside the cylinder. The air piston can overcome the spring pressure and clamp; and, when we wish to unclamp, we simply exhaust the air and let the spring do its work.

Although the spring force may not be as powerful as the air force, this arrangement allows the control mechanism to be much simpler. If there is an unexpected loss of air pressure during the use of spring-loaded cylinders, it is possible to determine exactly to what position the robot will return. If the air pressure is lost, and if we do not wish the robot to remain accidently gripped on something, this type of spring cylinder is able to release the gripper.

The use of springs for a robot with a long stroke is rare. When a long spring is compressed, the amount of force necessary to produce one small amount of motion at the *end* of travel, with full spring compression, is much greater than the amount of force necessary to produce the same motion at the *beginning* of travel. For this reason, a much larger-diameter air cylinder would be needed to overcome the spring force to accomplish the same task. The larger—and more expensive—air cylinder may add more to the cost of the robot than an air cylinder which operates in both directions.

### Hydraulic Power Supplies

In a hydraulically driven robot, the source of hydraulic pressure is usually a self-contained hydraulic power supply. The supply is assembled from several standard components: a hydraulic pump usually driven by an electric motor, a relief valve allowing excess pressure to escape back to the tank, the tank itself containing a supply of hydraulic oil, and often a tooling unit to keep the oil from overheating. The self-contained hydraulic unit may be placed some distance from the robot. This allows the space taken up by the robot to be smaller, and makes the robot more efficient.

If a robot has large pistons or must move very rapidly, a large hydraulic power supply is necessary. If the robot is small, or if the moves are done slowly—requiring a small amount of oil—a smaller hydraulic oil supply may be used. Or, one larger supply for several robots may be used.

In an air-driven robot, there is usually not an air compressing unit. Most manufacturing plants have a centralized air compressor. This large air compressor supplies air pressure used by a variety of equipment. Because this equipment is usually in place when the robot is installed, the overall cost of a pneumatic robot can be less than a comparative robot driven hydraulically.

By comparison, air is a more efficient, economic source of power than is hydraulics. When a pneumatic robot is standing still, the air compressor stores its current output of compressed air in a tank. When this supply tank is filled to the proper pressure, the compressor itself stops. If there is no use for this air for a long period, no energy is consumed. Air pressure is available at a moment's notice. At some future time, when plant equipment is being used sufficiently to remove air from the tank, the tank pressure will drop and the air compressor will start again.

In a comparable hydraulic unit, when the hydraulic oil is not needed to run the robot, the oil is difficult to store. It is possible to have a tank to contain pressurized hydraulic fluid, but this tank will have a limited supply of fluid. It is common practice that while a robot is operating, so is its hydraulic power supply. The motor turns the pump, and the pump pressurizes the oil. If the oil is not used immediately by the robot, it is allowed to pass through the relief valve and back into the tank (Figure 3.11).

The energy supplied to the electric motor, when not used by the robot,

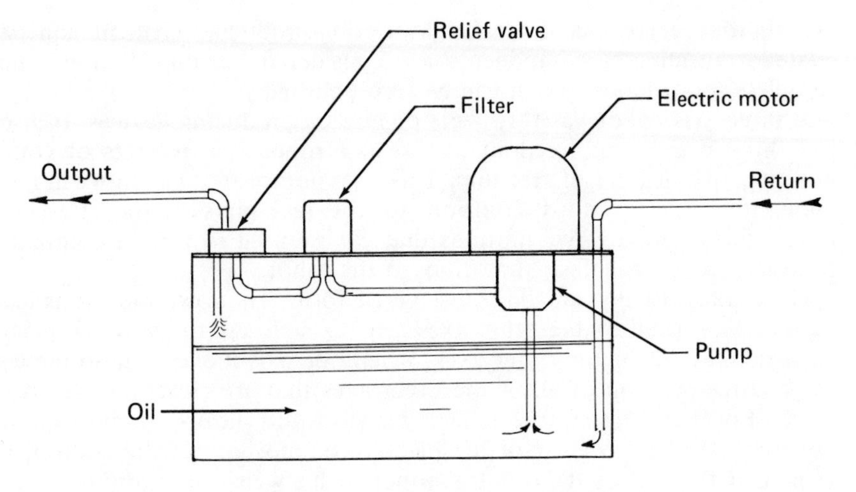


**Figure 3.11.** A hydraulic power supply.

is changed by the hydraulic supply into heat and noise. If the robot is working on a fast cycle or there are several robots using the same power supply, overall efficiency can be high. If the robot must work long periods when there is little or no motion of the robot, the overall economic efficiency of a hydraulic robot is low.

The actual power consumption of a hydraulic robot can add several thousand dollars per year in energy costs to the user. When hydraulic and pneumatic robots are compared to electrical robots, the same efficiencies can be seen. In an electric robot, power is used only when robot motion is taking place, and little or no electrical current is necessary to hold the robot in position.

In a sense, all three types of robots work on electrical power, the electrical robot receiving its power directly from electricity, the hydraulic and pneumatic robots receiving their power from an electric motor pressurizing some type of fluid. The energy efficiency of the air compressor or the hydraulic unit must be taken into consideration when evaluating the overall energy efficiency of the robot system.

## HOW SOME MOTIONS ARE COMBINED

The individual components comprising a pick-and-place robot are combined in various ways to produce different motions. It is common to use the term *axis*, or degree of freedom, as a means of comparing the abilities of robots. Each single motion-producing element and its related transmission components are considered an axis, that is, one complete set of motions that can be accomplished.

A three-axis robot has three sets of motion-producing devices. It also has three degrees of freedom. A five-axis robot has five sets of components, five degrees of freedom. This does not mean it can move in five dimensions—it still must conform to the real physical universe—but rather that there are five motions that can be used in various combinations to produce the desired motions of the robot.

These axes usually are built in layered form. The lowermost axis has mounted on it all of the other axes. The axis closest to the work piece is carried by all of the other axes. When one axis moves, it also moves the relative position of all of the other axes that are closer to the work piece. For this reason, the base is usually quite heavy. It must carry not only all of the weight of the piece to be moved, but the additional weight of the rest of the robot. A robot with a weight limitation of even a few pounds may have a base capable of moving 50 pounds, because of the weight of the rest of the robot.

When a robot designer contemplates the use of a robot, a consideration must be made for the proper, or most beneficial combinations of motions for the particular function. When using a pick-and-place robot, this is a matter of selecting the basic components to be used in each of the desired motions. Are the motions to be cylindrical? Should the base rotate or slide? How far must the arms reach? The designer must choose.

## ACCURACY

The accuracy of a pick-and-place robot is determined primarily by the tolerance of the manufactured components. When a cylinder travels as far as it can go, there is a definite position it will reach. This position will not change, or will change only slightly as the components wear. If the connections between the axes are tight, a pick-and-place robot will have a very high accuracy. That is, each time it does the same operation, it will return to exactly the same place.

It is common in the robot industry to see accuracies of 0.002 inches. This accuracy will remain stable over a long time, but should not be confused with the ability of the robot to reach any particular point. The end position the robot finds may or may not be the correct position for the job to be done. As the robot is installed, it is usually adjusted so the final position will be correct for the given operation. This adjustment, because it is mechanical in a pick-and-place robot, is sometimes inaccurate. Indeed, sometimes it is easier for the robot installer to move the equipment on which the robot is to work than to adjust the robot to its proper position. The robot controller has little effect in the accuracy of the pick-and-place robot.

Figure 3.12 is an example of a pick-and-place robot. This particular robot has the task of unloading a press. Each time the press makes one cycle, it will stamp out a piece of metal. The robot will sense the press to be in its proper position. It will reach into the press, grasp the metal, move it over to the basket, then release its grip. Control for the robot in this instance can be as simple as having its valves wired to the press controller.

There are two axes needed for this job. One axis will move the robot arm in and out. The other axis will control the grip functions, closed or open. (Because a robot is mass produced and may be used for many

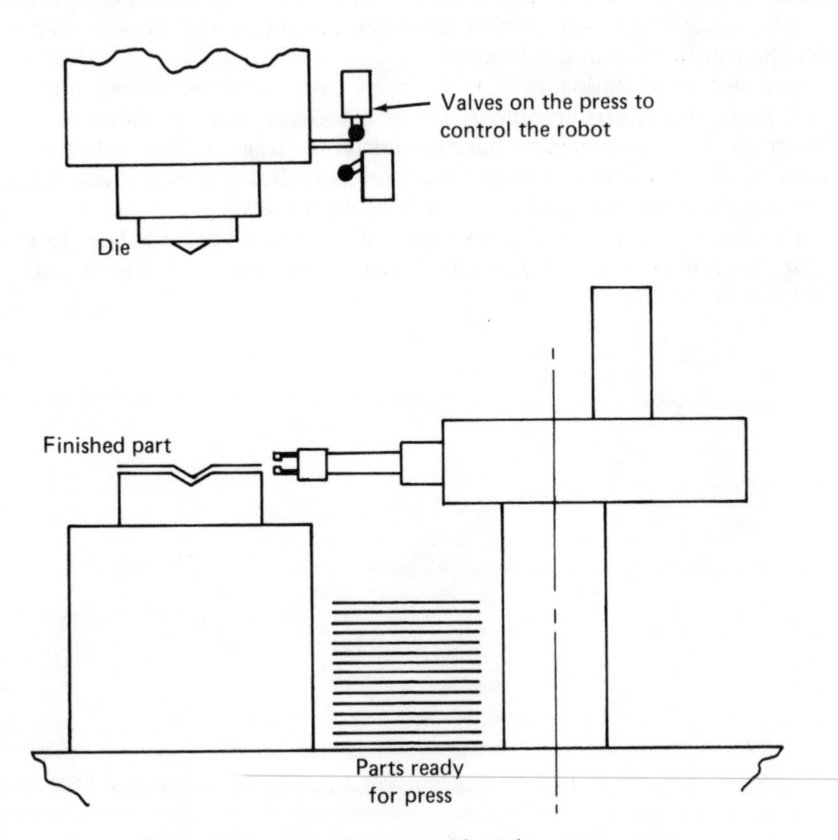


**Figure 3.12.** A robot actuated by valves on a press.

functions, the grippers are often added last and usually are not counted as an axis of the robot.) The sequence the robot will follow is always the same: extend, grip, retract, release. Providing there are no intermediary signals which would cause the gripper to be closed before it goes into the press, we need only repeat this cycle over and over to have a functioning pick-and-place robot. If the basket were located other than on a straight line with the press, we would need a three-axis robot to accomplish this same task.

Press loading is one of the primary tasks for pick-and-place robots. Imagine, instead, it is a human unloading the press. One after another, the operator removes parts, hour after hour. The tedium of this work function could cause carelessness. A tragic accident could occur if a hand or a finger were placed in the wrong place during the press cycle.

Many manufacturing plants use a series of safety stratagems to protect the worker. Buttons that require both hands to activate the press are installed in a safe location. When the press cycle is finished, the buttons can be released and the worker can then reach into the motionless press for the finished metal product.

A robot press unloader can work as fast as the press can operate, and the robot controller allows the arm to enter only when all is clear. If an accident does occur, there would be no tragedy. The robot is not human, it does not feel pain. The company has lost some equipment but a human operator still has the use of both arms.

The last chapter shows an example of a pick-and-place robot. In that unit, several of the components listed above are used, along with a simple control unit.

# CHAPTER 4

# SOPHISTICATED ROBOTS ARE STILL SIMPLE

The most sophisticated robots in use today are still simple to understand. Like the pick-and-place robots, they are assembled from components that are used for other functions in industry. It is the interaction of components and the type of components used that make the difference between the simple and the more sophisticated robot.

The components used to move the robot fall into three general categories: those used to produce motion, those transmitting motion, and those that limit motion. The control aspects of the robot and the feedback features will be discussed in a later chapter.

## MOTION-PRODUCING ELEMENTS

The most common components that produce motion in a sophisticated robot are *servo motors*, *stepper motors*, and *servo valves* used in conjunction with some type of *cylinder*, or *fluid motor*.

### Servo Motors

Servo motors are a special type of electric motor. Once supplied with electrical power, they rotate and can transmit energy to a robot component or mechanical linkage. What makes them unique is their ability to have controlled speed or position. With the aid of a control mechanism, a servo motor may be commanded to turn in either direction at varying revolutions per minute (RPM). A servo motor may rotate clockwise at several thousand RPM, and then quickly decelerate to turn for a few more moments at some low RPM. One type of servo motor can feed back to the control mechanism an indication of its relative position in rotation and cause the control mechanism to alter the motor's speed and path, to direct it to an exact position.

In robots, because they are concerned with accuracy, the non-feedback type of servo motor is generally used. It is enough for a robot controller to exercise its control over the speed and direction of the servo motor. A separate feedback loop can be employed to check accurately on the position of the motor, or, more importantly, the position of the robot arm itself. This separate channel, although creating a new series of equipment to be monitored by the computer, can be more accurate and less subject to failure than a servo motor mechanism that will accomplish the same task.

If we examine the signals given to a servo motor (of the speed-control type), we usually will see a variable voltage input as a signal to the motor of speed. When the voltage is low, a low RPM is produced. As the voltage is increased, the motor speed increases. The control mechanism may have a simple look-up table, or digital analog converter, allowing very fine control over the servo motor speed.

Elements of acceleration and deceleration also may be controlled directly by the computer with the aid of the external monitoring device. As a servo motor achieves some position for the robot and this position must be maintained, the motor's ability to revolve very slowly can be of great help to the computer. As the position of the robot would begin to drift from its desired point, the computer would command the servo motor to turn slowly and thereby correct the position. In the case of a non-servo motor—where the motor would revolve at full speed, once accelerated—fine adjustments in position and the stillness needed in some robot operations are nearly impossible. A servo motor can produce very smooth motions without the jerks associated with some other components.

## Stepper Motors

A stepper motor is also a specific type of electric motor. The windings within a stepper motor, however, are not placed so as to continuously revolve a shaft. Rather, they are a series of electromagnets that will alternately, and in pairs, pull a flexible gear component into position. Figure 4.1 shows the basic components of a stepper motor. There are a series of electromagnets, a flexible gear component, a circular gear component, and the necessary bearings to keep each of these elements in their respective positions. Motion on the stepper motor shaft is produced by a differential in the number of teeth in the flexible and the circular gear components. As a pair of electromagnets on opposite sides of the shaft are energized, the flexible gear is brought into intimate contact at those points with the circular gear. As long as the gears remain in contact under the influence of the electromagnets, the stepper motor will achieve its holding torque.

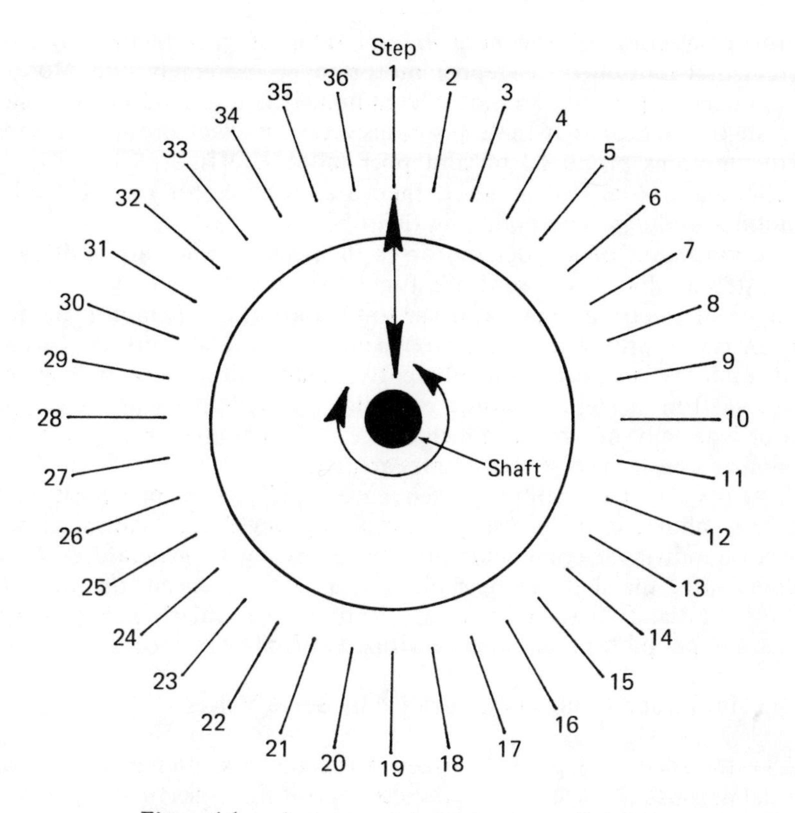


**Figure 4.1.** A stepper motor moves in small steps.

The ability of the shaft to resist turning in either direction is limited by the mechanical meshing of the gears. When motion is required on the shaft, the next series of electromagnets will be energized. As the flexible gear and circular gear are brought into contact, power will be released on the previous set of electromagnets. Because the differential of the two gear components is very small, when any one pair of electromagnets has engaged the gears, the pairs on either side of them will have gear teeth very nearly engaged.

As gear teeth are continuously engaged in a rotary fashion around the stepper motor, each set extends slightly the position of the circular gear. Examining the relative position of the gear teeth located 90 degrees from those in action shows us that the individual teeth of the two components will move from one mating number to another mating number in one-half rotation of the motor shaft. It is by this continuously propagating wave that the stepper motor produces motion on the shaft.

Proper selection of the number and ratio of gear teeth allows the incremental motion of a stepper motor to be extremely fine. Motions of 1.0 degrees per step are not uncommon. It is a natural phenomenon in a stepper motor that these positions are very exact in their rotation. If the motions produced by a stepper motor are fine enough for the positioning accuracy of a robot, then a stepper motor can be used as a motion-producing component in the robot.

One limitation of stepper motors is their low power output. Because each step is done as a separate function, only the energy developed by a set of electromagnets is of use to the output of the motor. Because the energy is provided by electromagnetic force and, further, because that force is produced by inherently small coils, the actual energy available from a stepper motor is small. Torque waiting for a stepper motor is usually measured in inch ounces, whereas in a servo-type motor of similar size, it may be many times greater.

A stepper motor's ability to sequence rapidly from one position to the next allows it to be used as a rapid-turning device. Although hundreds of individual commands may be necessary to generate even one revolution of the shaft, stepper motors can achieve speeds on the order of one thousand RPM. In many applications, particularly involving light loads, stepper motors can be used with great facility in robots.

### Fluid Motors and Cylinders Controlled by Servo Valves

Like the special-purpose electric motor called a servo motor, there are special-purpose fluid components also operating as servo devices. It is not the components themselves that must be different, but the valve mechanism that controls the flow of fluid to the components. A general-purpose hydraulic cylinder can be used with a servo valve to produce a servo-controlled hydraulic component. A servo valve is a special type of flow control valve where a variable input signal will produce a variation in the flow of oil passing through the valve (Figure 4.2). Air is usually not controlled by a servo valve.[1]

---

[1]A special type of servo valve has been invented for use with an air piston. Rather than allowing the valve to open and close only as motion is desired, this air valve allows the air to energize continuously both sides of the piston. The valve itself is constantly in motion, alternately charging one side and then the other in the piston. Valve speeds on the order of 1,000 cycles per second must be used. By very sophisticated feedback mechanisms, the exact position of the piston can be maintained by slightly altering the time the moving valve will spend in any one particular position. With the aid of the computer, these very high-speed oscillations are capable of overcoming the weakness of air as a servo medium.

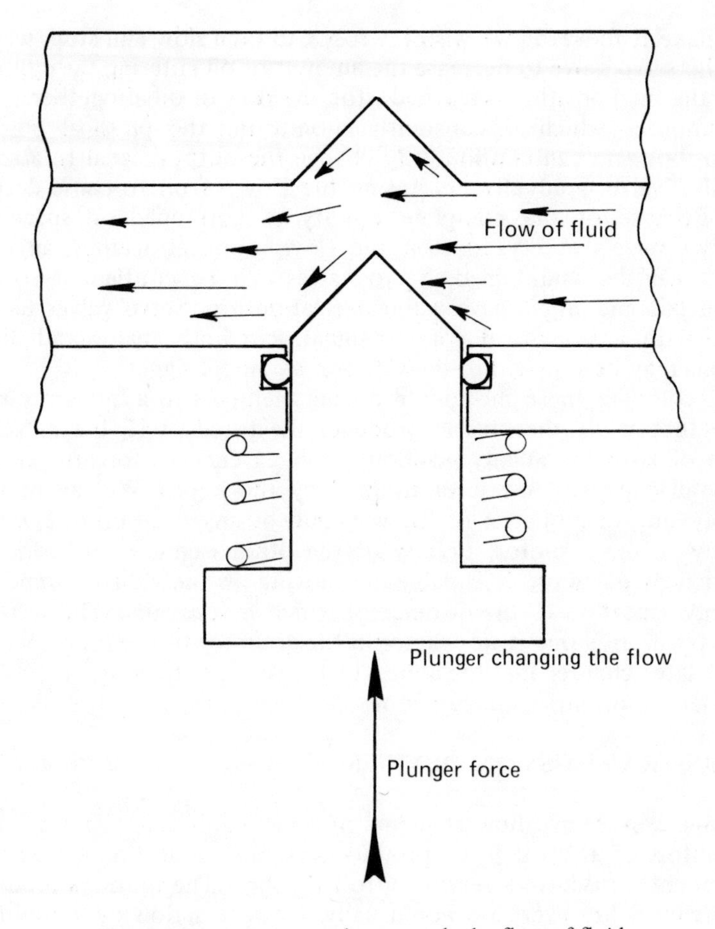


**Figure 4.2.** A servo valve controls the flow of fluid.

There are servo valves for both position control and speed control of hydraulic components. When a valve mechanism is configured for a position control, a speed-back position is established between the physical position of the moving member and the control mechanism of the servo valve. Such arrangements, because they do not allow for intervention by a robot's computer, are usually not used on robots. Instead, it is a servo valve–with its ability to control the speed of a hydraulic component–that is most often used.

If we wish an arm of our robot controlled by a hydraulic cylinder to move quickly, we are able to supply large amounts of pressurized oil to make it move. If we wish the robot to then slow and stop, we will cause the servo valve to decrease the amount of oil entering the cylinder, and, as the final position is reached, stop the flow of oil altogether.

A computer which is constantly monitoring the physical position of the robot arm can continuously change the output signal to a servo-controlled valve, gradually decreasing the flow of oil to cause deceleration. Because of the complete rigidity of a trapped oil space and the servo valve's ability to stop the flow of oil altogether, an exact position can be maintained. Servo valves can be configured so that their output and input have a linear relationship. Servo valves usually operate with an electrical control signal, and both analog and digital functions may be supplied to the valves as a control signal.

If we compare these motion-producing elements to a human muscle, we see that while the muscle produces the force itself, the muscle is capable of stopping at any position. If we extend our forearm, it need not go all the way out until stopped by the bones. We can move it half way out, or a quarter of the way out, or any position in between. Similarly, a servo motor, or any of the other elements of motion in a sophisticated robot, is capable of moving an individual component over only a portion of the distance of which it is capable. The accuracy of this position is one major factor in the accuracy of the robot. We will see in a later chapter that the ability of the robot to measure its position is another important accuracy factor.

## MOTION-TRANSMISSION DEVICES

Chapter 3 explained how elements of a pick-and-place robot are built one on top of the other to produce a series of motions. This same arrangement is used in a servo-controlled robot. The motions produced by the robot, however, are not usually a direct action of the motion-producing devices. Rather, as the motion is produced, it is transformed from its natural motion into one more useful for the robot in its task. There are a few common transmission elements used in the majority of robots. These are: the *ball screw*, the *chain drive*, the *harmonic drive*, and *direct mechanical linkages*.

### The Ball Screw

A ball screw is quite similar to a screw and a nut (Figure 4.3). There is a screw portion that has a groove around its outer edge. This groove is one continuous path around the outside of the cylinder. The nut portion of the ball screw is a box containing a series of ball bearings. The box is called a cage and is used to transmit the weight placed on it through the ball bearings, onto the screw. The cage also cycles the ball

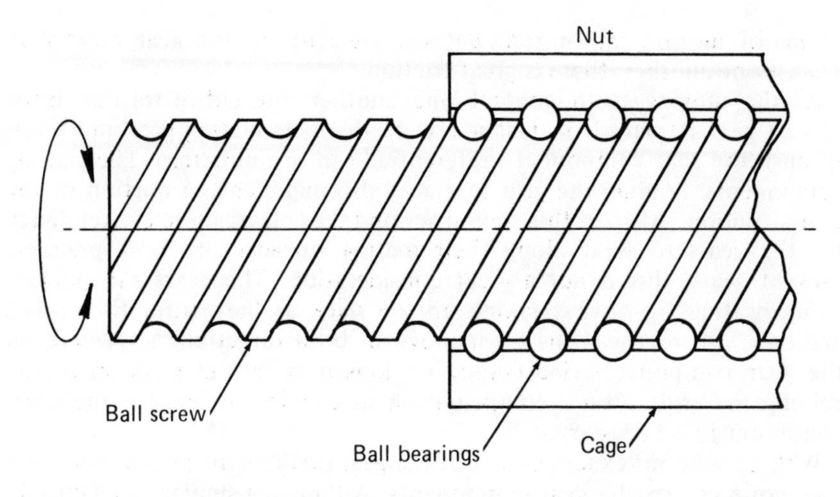


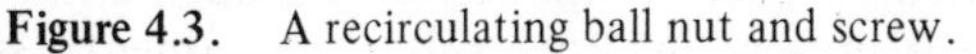
**Figure 4.3.** A recirculating ball nut and screw.

bearings through a continuous path allowing the limited number of balls to engage continuously on the threaded portion of the screw.

As the balls in the cage roll against the screw, they come to the end of their travel. Each ball in turn goes through a special groove arrangement in the cage, travels backward through the cage, and starts again at the beginning of the rolling motion. In this way, each ball is at times under load against the screw and cage, and, at other times, is free from load and travels to begin its job again.

A limited number of balls, continuously rolling against the cage and the screw, can be used to travel a theoretically infinite distance along the screw itself. There is no time and no position in which a minimum number of balls is not engaged to carry the load Because there is no limit to these positions, a ball screw is a very efficient method of transmission. The power necessary to turn the screw shaft is almost equal to the power that will be generated by the moving cage against the arm of the robot.

## Gear Drives

Regular gear components can be used to transmit motion from one part of a robot to another. However, there are several reasons why gear components are used only in the most simple of robots. These relate to standard gears' mechanical efficiency, wear characteristics, size, and noise level. The typical involute tooth gear is not a mechanically efficient means of motion transmittal. Because elements of the gear must slide against one another, there is great friction.

As the moving teeth contact one another, the entire torque thrust of one gear is placed on just a few teeth of its mating gear and tends to push the gear component preferentially in one direction. The bearing components, holding the gear in place, also experience a portion of the gear's friction. Also, as this sliding action takes place, there is a tendency for the gears to wear along their mating surfaces. In most practical uses of gears, this is not a great consideration. The gears can be self-compensating by always taking up the slack as they turn. In a robot however, where the arms must move in both directions anywhere on the gear components, inaccuracy or looseness in the position of the robot can result. Some compensation mechanism to engage the gears continuously is necessary.

With or without exact contact of the gear surfaces, there is an unavoidable noise created by gear components. Although a similar mechanically produced noise is present in other robot components, such as harmonic drives and chain linkages, the noise is greatest in regular gears. This is because a normal spur gear is much larger than the other components which produce a similar motion. Although the industrial environment is not noted for its quietness, even a small amount of extra noise in the form of a noisy robot is strongly discouraged by most robot customers. The esthetics of the work place are not as much a consideration as the belief that a noisy machine is also one subject to failure.

All of these comments do not mean that the spur gear cannot be and is not used as a robotic component. Rather, it is used rarely and only after close scrutiny of the effect on the robot's performance.

## Harmonic Drive

The harmonic drive is a special type of gear component extensively used in robots. Although an American invention, it is not produced in America. Because of the high precision required in the creation of a harmonic drive and the expensive machinery employed, it is not considered profitable to manufacture harmonic drives in the United States. Oddly enough, the use of robots can reduce manufacturing costs sufficiently so that these robot components themselves might be domestically produced.

The harmonic drive (Figure 4.4) consists of three components. There is a roller bearing shaped as an ellipse called the *eccentric bearing*, a flexible gear canister with very fine teeth ground on its outside surface called the *flex spline*, and a *circle spline* with teeth on its inside. The circle spline has teeth very similar to those on the flex spline; but, in order for a harmonic drive to function, there must be a different

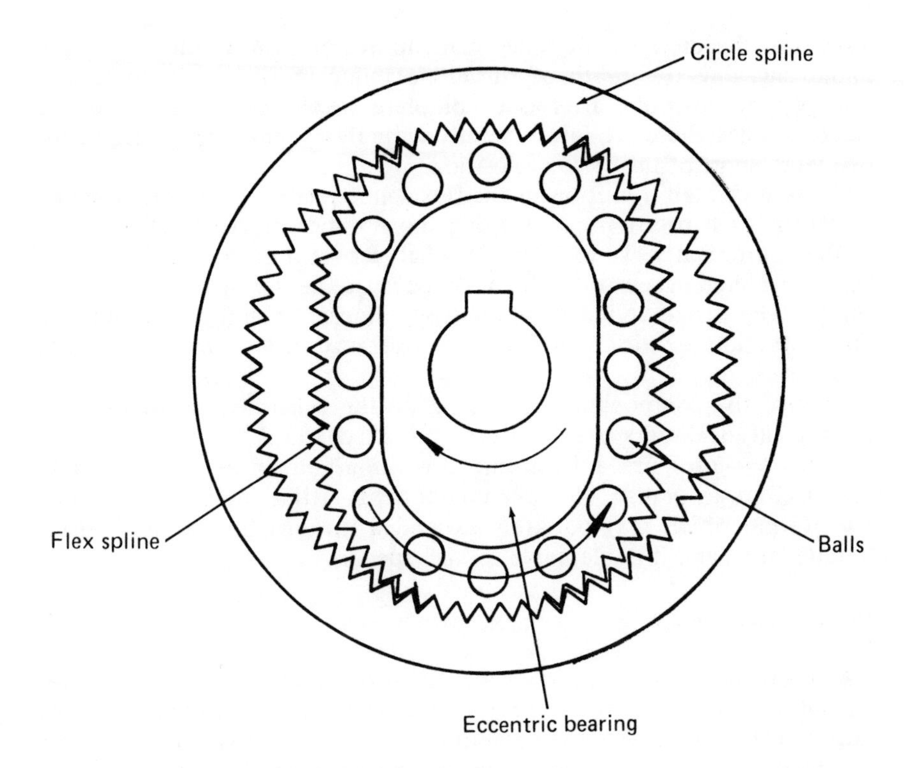


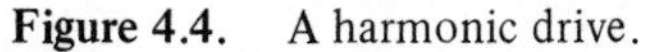
**Figure 4.4.** A harmonic drive.

number of teeth on the two spline components. We will use the example of 200 teeth on the flex spline and 202 teeth on the circle spline.

The components are placed one inside another. The eccentric bearing fits inside the open portion of the flex spline. The flex spline is placed inside the circle spline so the two sets of teeth are in line.

As some component producing motion, such as a servo motor, revolves the eccentric bearing, there is a peculiar effect in the engagement of the gear teeth. Because the eccentric bearing is sized so that its long side will cause snug contact in the two sets of gear teeth, we see that there is clearance between the gear teeth along the short side of the bearing. As the bearing rotates, this position of engagement and disengagement travels around the circle spline. First, a portion just at the top and bottom is engaged, then a portion more to the right, still more to the right, at the sides, slightly below the side on the right, and so forth, until the original engagement rotates 360 degrees.

Because the engagement takes place uniformly, the tooth of the flex spline engaging the tooth of the circle spline most near it, there is a disparity of positions after one complete revolution of the eccentric bearing. Since there are fewer teeth on the flex spline, the rotation ends two teeth short of the original position.

We have created a motion in the flex spline relative to the circle spline of 1/100th of a rotation. By varying the number of gear teeth in each of the components, we can achieve whatever ratio of mechanical advantage considered necessary to our task. Because the flex spline component engages the circle spline in a relatively linear fashion, rather than a sliding fashion as with a spur gear, we see very little wear on the teeth themselves.

Because the components fit one inside the other, we need not take up a great deal of space to create this mechanical transmission. It is possible to configure a harmonic drive component in approximately the space of an orange, all three components fitting inside that volume. The torque which may be safely expended through the harmonic drive is quite large, and it is adaptable to high speed.

## Mechanical Linkage

A mechanical linkage (Figure 4.5) is also sometimes used as a transmission device in a robot. The rotation of one disk causes the connecting rods to rotate another disk. Sometimes we have two rigid elements, one mounted on the motor shaft, and another fixed to the motor housing. A displacement between these two rigid elements works like a lever to produce motion in one of the robot arms. As with the pick-and-place robot, if the arm of the robot must go through some small opening to do its task, it is not necessary to have the motion-producing devices

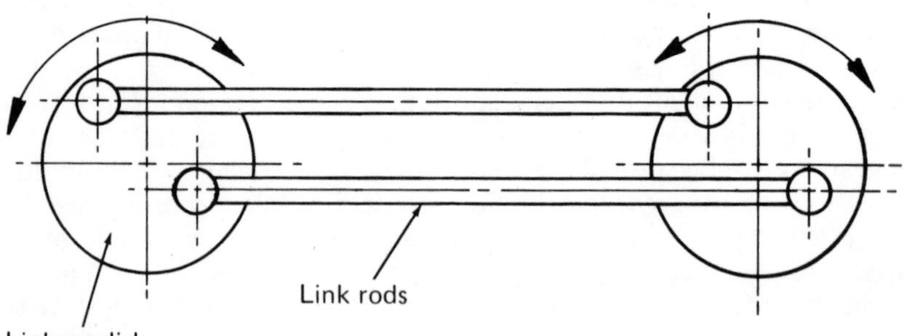


**Figure 4.5.** A mechanical linkage.

(which can be very large) at the "working end" of the robot. Rather, they can be on the robot base, with only the mechanical linkage element out on the end. As a sophisticated robot is being designed, combinations of motions must be selected to best suit the robot's work.

## MOTION-LIMITING DEVICES

In addition to producing motion to be used in a robot arm and transforming that motion to its most useful state, a servo-type robot often uses devices that limit or restrict the motion of its arms. Elements commonly used for this purpose are *mechanical brakes, shock absorbers, locking valves, shot pins* and *collet-type motion limiters.*

### Brakes

The brakes used on robots (Figure 4.6) are similar to the caliper-type brakes used on automobiles. There is a disk of dense high-friction material which rotates with the moving component of the robot. Placed on both sides of this disk (or just on one side in some cases) are pads of the same type of material. When the brake is engaged, the pads contacting the rotating disk do not allow the disk to turn. If the disk is coupled directly to the arm of the robot, this limitation to motion can be only as good as the pads' ability to completely restrict motion. If the robot is to maintain accuracy under varying load conditions

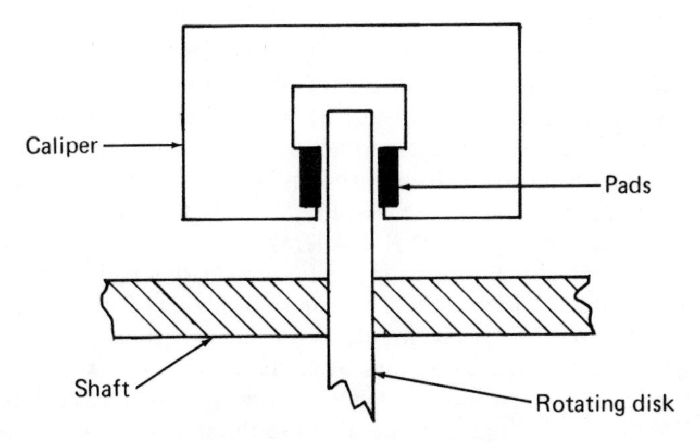


**Figure 4.6.** A caliper disk brake.

of only a few thousandths of an inch, the brakes must be extremely tight and have a large surface area.

It is common practice to couple the brakes, instead, to the motion-producing device, either servo motor or some other rotary equipment. The brake in this case limits the motion of the motor and, through the coupling of the transmission element, limits the motion of the robot arm. If the brake slips, the amount of motion produced in the robot arm is very small, since it is reduced by the ratio of the transmission component.

It is standard practice to have a brake component that is driven in place by a spring or magnetic force. By the use of a permanent magnet or spring, we can be assured that the brake always will be engaged should there be some control or power malfunction. Even if power to the robot is disrupted completely, the brake will automatically set. In these cases, power is used to unload the force of the brake. The robot computer, before it attempts any motion on the part of its motion-producing devices, will first stimulate the brakes to their release position.

### Shock Absorbers

Fluid-filled and air-cushioned shock absorbers are also used as standard robot components. Sometimes, as the arms of a robot are moving, they will have a tendency to move abruptly, for example, as the arm of a robot extends far off balance. If a shock absorber is placed to cushion this effect, a great overall accuracy can be achieved, for both speed and position. Where a robot may reach the end of its travel but still be moving at high speed, a shock absorber can rapidly decelerate the robot arm without causing physical damage.

The shock absorbers used in robots are similar to, but more sophisticated than, those used in automobiles. A typical fluid-filled shock has a piston traveling in a cylinder containing a series of oil-restricting orifices (Figure 4.7). As the plunger of the shock absorber is engaged, it causes the piston to force oil out of its path and through the orifices. Initially, the hole size is great enough so there is little force resistance. As the piston proceeds through the cylinder, it covers up the larger holes. Progressively, the force required to depress the plunger is increased, and the moving object is decelerated.

In one type of shock absorber, a silicone oil specially designed to be compressible under high loads is used in place of the conventional hydraulic oil. In this arrangement, a special orifice chamber produces very high velocity oilstreams and actually compresses the silicone oil. The net effect to the robot arm is the same in either case. What could be a catastrophic striking of a metal component against one of the other robot surfaces is instead limited to a controlled exchange of force.

In some instances, an air-filled chamber is used as a shock absorber.

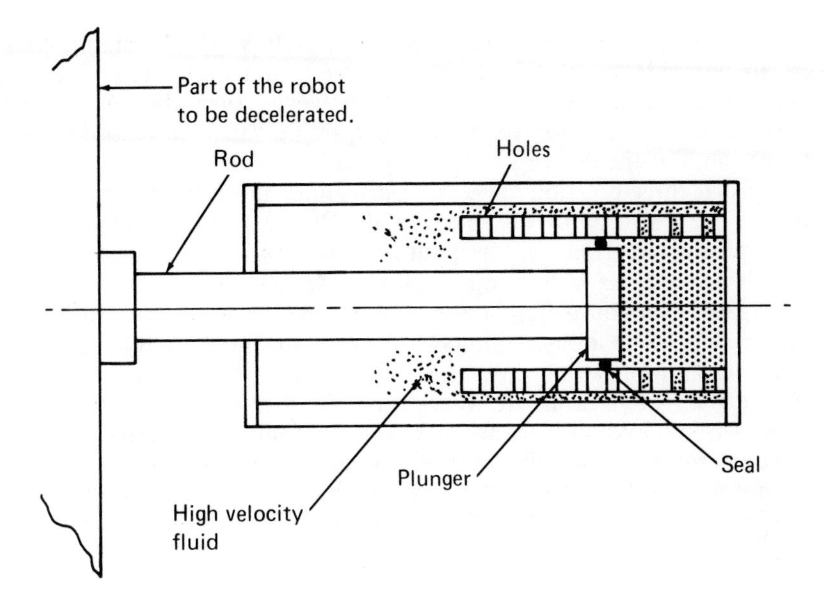


**Figure 4.7.** A hydraulic shock absorber.

The compressibility of air, particularly when sealed by a rubber bladder, can be used to cushion the motion of robots, and, in some cases, limit their travel. If an air cylinder is used as a component of a robot, an air shock absorber can be built directly into the internal portion of the air cylinder. (This arrangement is sometimes used in hydraulic cylinders, but less frequently on those controlled by a servo component.)

As the air piston proceeds toward its maximum extension, the thickened section of the piston rod will enter a pocket in the air cylinder's cap. Until this entry, the exhausting air escapes through an opening within the pocket and through the valve. Once this rod section enters the pocket, however, the flow of exhausted air is limited to that which may pass through a small hole or groove adjacent to the pocket. This cushion of air at the end of the cylinder's travel is considered a shock absorber. When we wish to reverse the motion of the cylinder, a special check valve can allow a larger volume of air to flow in the opposite direction, and quickly return the piston to the other end of the cylinder.

## Locking Valves

A special valve arrangement is sometimes used in hydraulic robots allowing the oil within a hydraulic component to act as a motion-limiting device. In simple valving, we alternately supply hydraulic pressure to one side or another of the motion-producing device. When we wish the device to remain in position, however, even though we may apply some varying load to it, the flow of oil must be restricted out of both sides of the device.

A special three-position valve is used to limit motion. The three positions of the valve allow oil to move the device in one direction or in the opposite direction. Or the path of the oil can be sealed altogether. In a device such as a hydraulic cylinder, where the oil on both sides of the piston is completely trapped, the position of the piston is held quite rigidly. The non-compressible nature of the hydraulic medium allows no motion whatsoever.

When the robot arms are very heavy in themselves and are being positioned by hydraulic components, it is common practice to have a locking valve which will return to the locking position whenever it is not supplied with power. This insures that a power failure will not cause the robot's arms to collapse catastrophically due to their own weight.

## Shot Pins

In certain instances, it is necessary to lock a robot's arms into position. There may be a certain point such as the home position where we wish the robot's ability to move to be completely overcome. And, we wish to force, by some external means, the arms of the robot into an exact position. This is commonly accomplished by a shot pin (Figure 4.8).

A shot pin is a mechanism similar to the bolt in a door. To engage the pin, a fluid piston is indexed to force the pin into a close-fitting hole. Unlike a door bolt, however, this pin fits so closely into the hole, no motion is allowed in any direction.

Further, the pin is tapered so that as it begins to engage, it can, if sufficient force is applied, overcome the robot's own mechanical components and force it into the correct position. Once shot pins have been engaged, it is possible to remove robot components and allow power to be shut off.

If each access of the robot is restricted by the shot pin engagement, shipping the robot, or storing it for some time, will not allow the arms to drift in their position relative to the presumed position of the control cabinet. In some robots, this shot pin engagement is an almost daily task to insure the exact positioning of the robot at its home position.

## Collet Type Brakes

The term collet refers to a loose-fitting sleeve which may be tightened around the shaft. Collets are used in many places in industry, especially

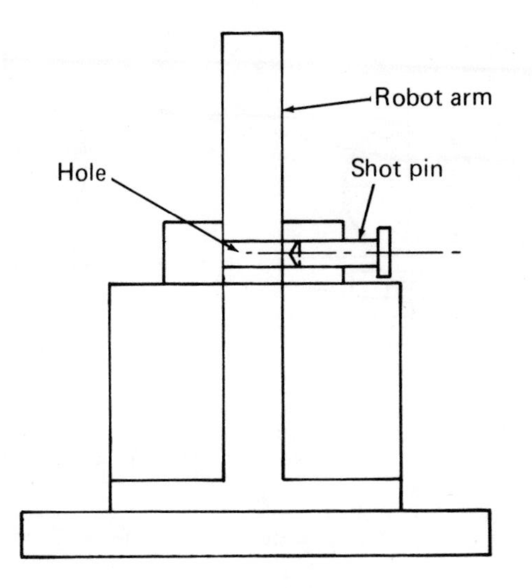


**Figure 4.8.** A shot pin used to fix the position of a robot arm.

where a cylindrical tool must be joined to some type of machine. In robots, a collet activated by some automatic equipment can be used as a brake mechanism. If one of the motion-producing devices is operating along a linear path, a collet can limit its motion. In some instances, by timing the engagement of a collet, we can use it to position an otherwise non-servo component. For example, a robot with extension-type arms can have a collet brake to limit the extension.

## SERVO LAG

One phenomenon occurring in servo-controlled robots is known as servo lag (Figure 4.9). It is a manifestation of a motion-producing component's inability to start or stop instantaneously. As a command is given from the robot computer to begin motion, we could, with very fast test equipment, measure the actual time that passed until that motion was created in the servo component.

Upon acceleration, this servo lag is of little concern. When the same phenomenon exists when stopping the robot, some negative features of the component become evident. Should a particular robot program require that an arm be moved rapidly to a certain position and then stopped abruptly, we may see a servo lag cause the arm to overshoot

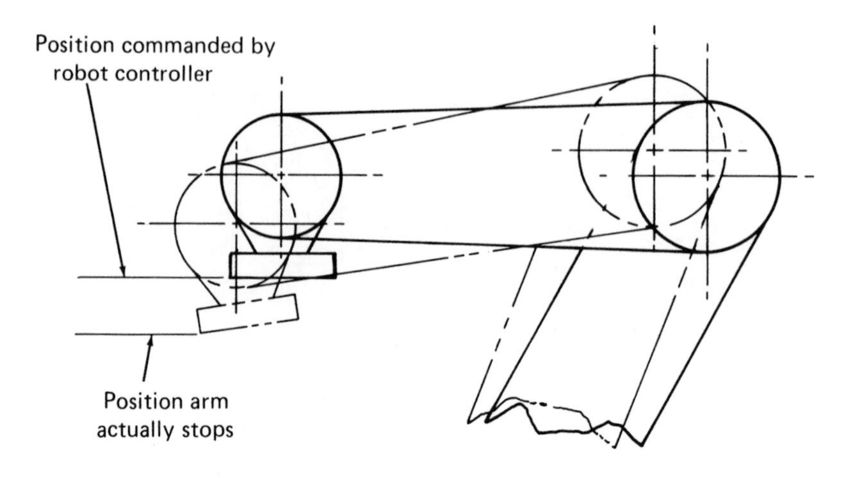


**Figure 4.9.** A robot arm showing the effects of servo lag.

its desired position, at least temporarily. When the robot is used with equipment such as vision systems or other sensing devices, servo lag may adversely affect the performance of the equipment. Robots performing process work, such as welding or grinding, should be designed for a minimum of servo lag.

## COMPLEX MOTION

One feature of servo-controlled robots is their ability to move all their axes through different portions of the individual arrangements (Figure 4.10). There is no simple way to describe the complex motions of a servo-controlled robot. With multiple arms all moving simultaneously, the path available for the end of the robot arm is almost limitless.

Many manufacturers include in their product literature a drawing showing the work envelope of the robot. This envelope shows the robot's ultimate reach when each of its axes is at the extreme of its motions. The envelope often will have multiple curved surfaces and appear irregular. While the end points of the work envelope may be easy to demonstrate, the actual motions needed by the individual arms to generate any particular path can require a great number of mathematical computations.

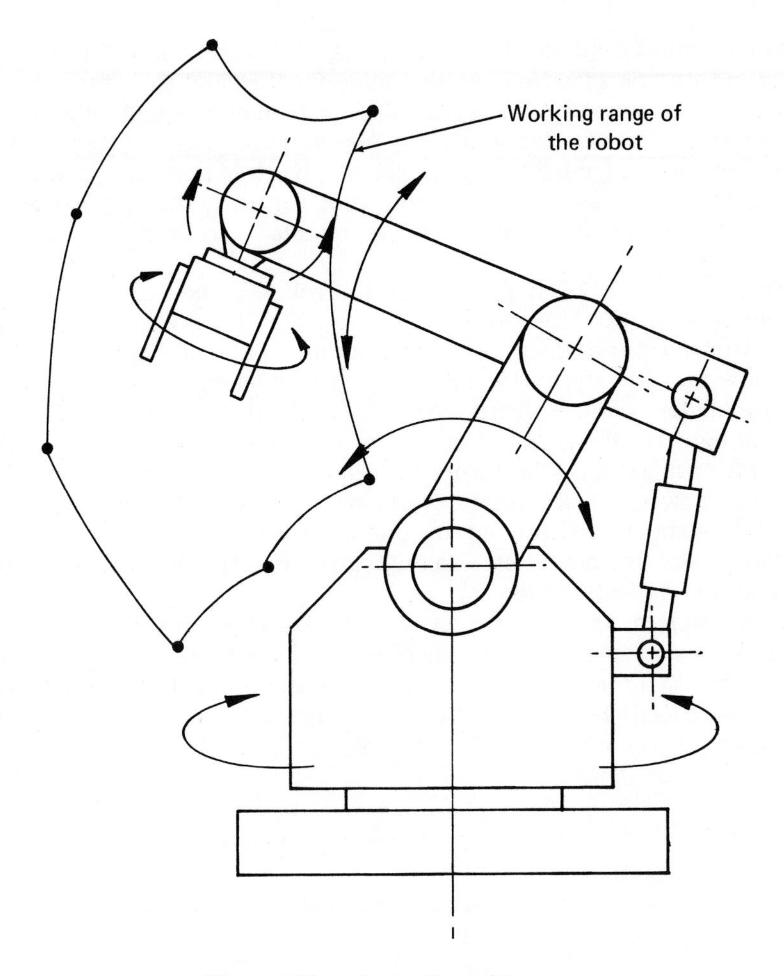


**Figure 4.10.** A robot's working range.

## ACCURACY

The accuracy of a servo-controlled robot is primarily the result of two parameters: one, the resolution of the components, and two, the care with which the physical components are built and assembled. The resolution of the motion-producing components is a factor of their ability to find and maintain a particular position in their path. The resolution of a servo motor is a phenomenon of that particular device.

As the component itself is designed, a minimum expected resolution is taken into consideration. As this component is used to design a robot, this minimum resolution dictates the approximate positioning accuracy of the robot. As the robot is built, however, if there are inaccuracies in the mechanical portion of the robot, how this portion is assembled and how all of these assemblies will work together, it is impossible for the motion-producing component to maintain accuracy in the arm of the robot.

Inaccuracy of a given position of the robot is the result of a combination of motions of more than one axis. If we were to compute by formula the accuracy of a given robot, we would wish each axis to know its accuracy by the resolution of its components and the physical parameters of its construction. We would then multiply these two factors by each of the other factors for each axis. A review of this type of formula might lead the novice to believe robots are grossly inaccurate devices. However, it is common practice to expect a servo-controlled robot to maintain an accuracy on the order of 0.01 inches. Some manufacturers will maintain an accuracy greater than this for particular functions in individual robots.

An example of the use of a servo-controlled robot is shown in Figure 4.11. We see an aerial view of the robot working in the center of a series of machines. There are several conveyers supplying assemblies to the robot work cell. Because this is a servo-controlled robot, it can go to

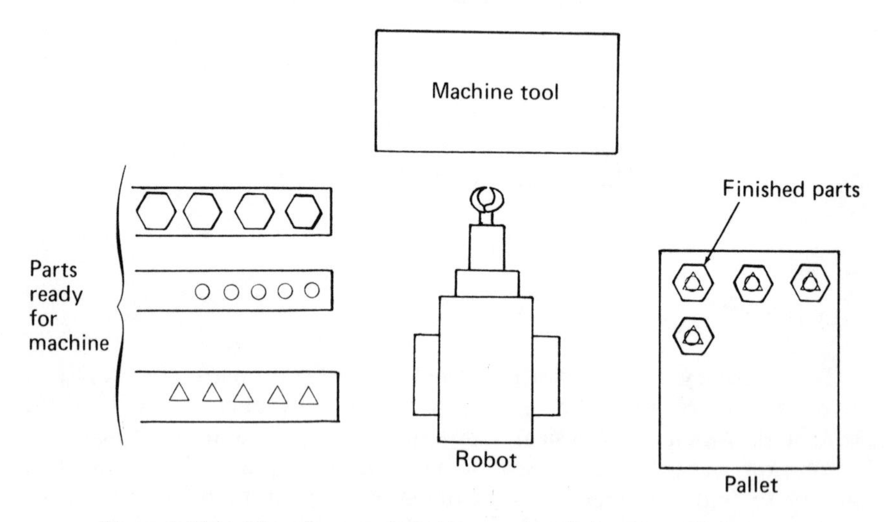


**Figure 4.11.** View from top of robot work cell showing palletizing.

each of the conveyer positions to pick up a part. These parts are placed sequentially in the machine. When all three parts are in the machine in the correct order, the machine will cycle and join the parts together as an assembly.

When the cycle is complete, the robot will remove the assembled components, take them to a grinding machine, and, when they are finished, place them on a pallet. The pallet in this case has nine different positions. The robot can sequentially place assembled components first upon position one, then upon position two, and so on. This is not possible with the pick-and-place robot. The number of positions to be reached would require a robot with incremental control of its axis. We can see, however, that the individual motions produced by the robot for each of these tasks are quite simple. To go to conveyer one, it is a simple matter to stimulate a servo motor to its correct position for the base, for the arms, and for the gripper. It is the combination of motions, however, that lets the robot do what we would have previously considered to be a human type task: loading the machine and then loading the parts on the pallet.

# CHAPTER 5

## SOME ROBOTS ARE HYBRIDS

While most robots can be considered either pick-and-place or servo controlled, there are some that have features of both. These robot hybrids can be used in applications too difficult for a simple pick-and-place robot, at less cost than a servo-controlled robot.

As a robot is being designed for a general type of work involving several functions, sometimes only a single task requires servo-control ability. If the number of positions to be reached by the robot is small, and yet one of these positions varies, it is possible to design a robot that would be a pick-and-place type with only one of its arms under servo control. Because the controller necessary for this type of robot would have all of the features of a servo-controlled robot, the cost would still be great, and perhaps the savings would not justify having only one axis servo controlled.

It is possible, also, to have a robot that will perform more complex tasks but have no servo-controlled functions. A robot of this type can use a multiple number of air cylinders, all producing the same general motion, but at different increments along the travel of the arm.

Consider an example where only a few extra positions are needed for one arm. Rather than having one air cylinder with a stroke of ten inches, our hybrid will have two small air cylinders with strokes of four and six inches. In a simple pick-and-place robot, the air cylinder would be at either position A, all the way back, or position B, ten inches out from position A. In our hybrid system using two air cylinders, as shown in Figure 5.1, we would have a choice of four positions: position W, both cylinders back; position X, only one air cylinder turned on; position Y, the other air cylinder turned on and the first one off; or position Z, both cylinders turned on. These positions would be measured zero inches, four inches, six inches, or ten inches.

It is possible to have three or four, or any number, air cylinders

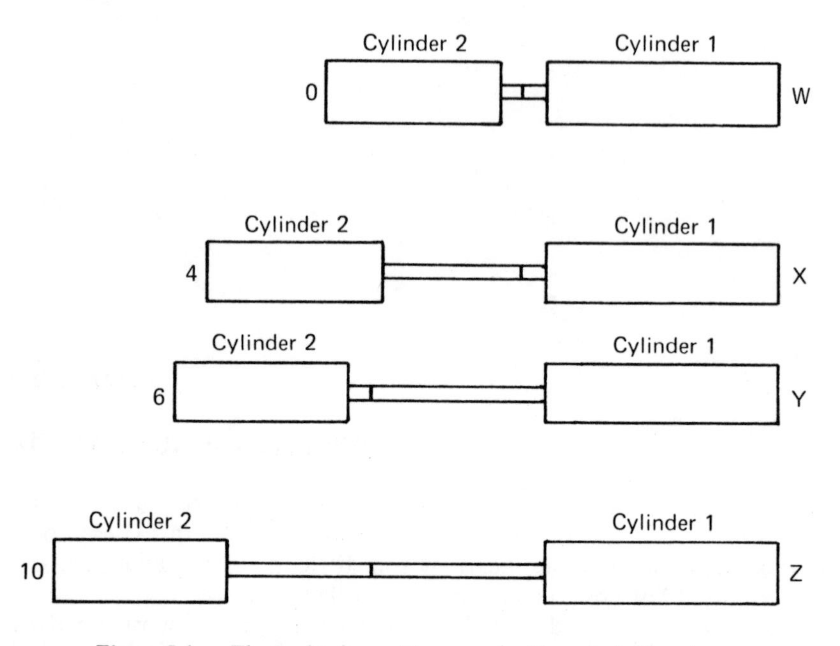


**Figure 5.1.** The multiple positions reached by a hybrid system.

connected end on end in this fashion to reach any finite number of points. However, as the positions increase, the control system necessary to activate each cylinder individually would become as sophisticated and costly as the control for a servo mechanism, therefore defeating our economic purpose. Within reason, however, it is possible to make an arrangement of components that will reproduce the motions of the servo-controlled robot without its cost.

In one viewpoint, our new multiple-cylinder robot is just a pick-and-place robot with more axes of motion. Several of the axes happen to be on the same centerline and moving in the same direction. To the robot programmer, however, there is a difference in the way the robot with multiple stop positions can be used.

Figure 5.2 shows a robot that must pick up a part and place it in a machine. When the machine has completed its task, the robot must place the same part in a parts basket. If the position of the input conveyer and the machine are fixed, there are only two positions on the floor that can be used for the parts basket if the robot is a traditional pick-and-place, three-axis machine. The only positions not used for other work are: in line with the *conveyer* with the arm all the way back, and in line with

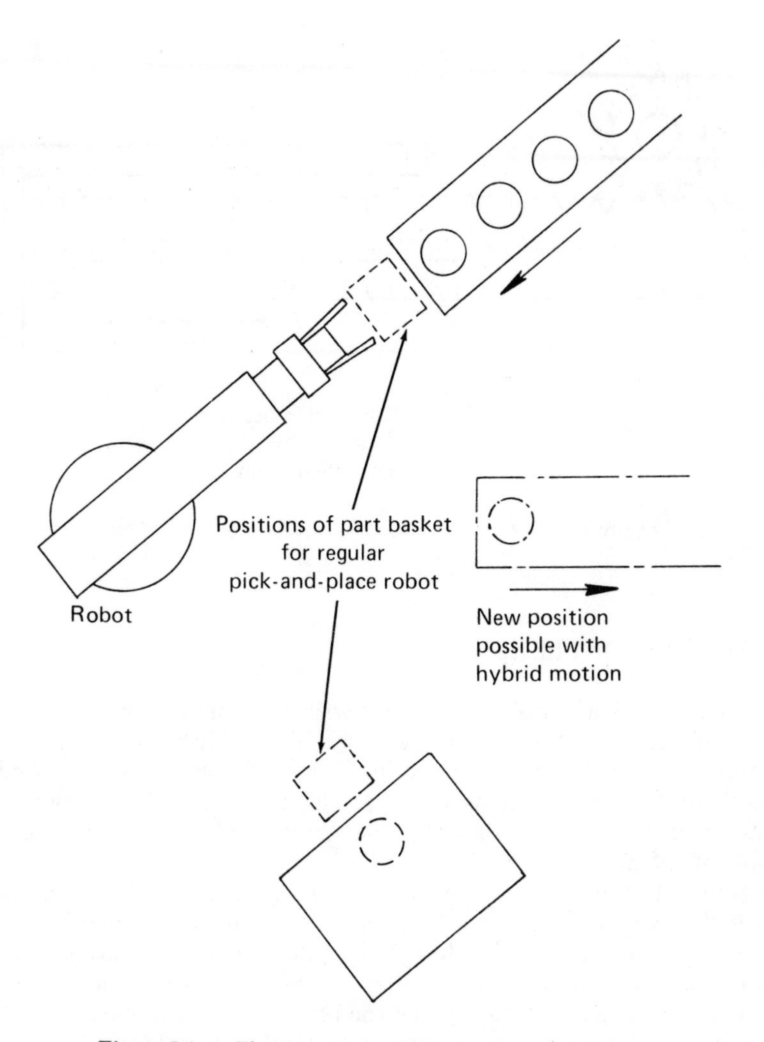


**Figure 5.2.** The increased usefulness of a hybrid robot.

the *machine* with the arm all the way back. If we wish to put the basket in another place, we must at least add another arm to the robot.

A hybrid robot with the same number of axes could have two pistons controlling the rotation of the base. Figure 5.3 shows the piston arrangement that might be used in the base. Another arm need not be added to increase the robot's usefulness.

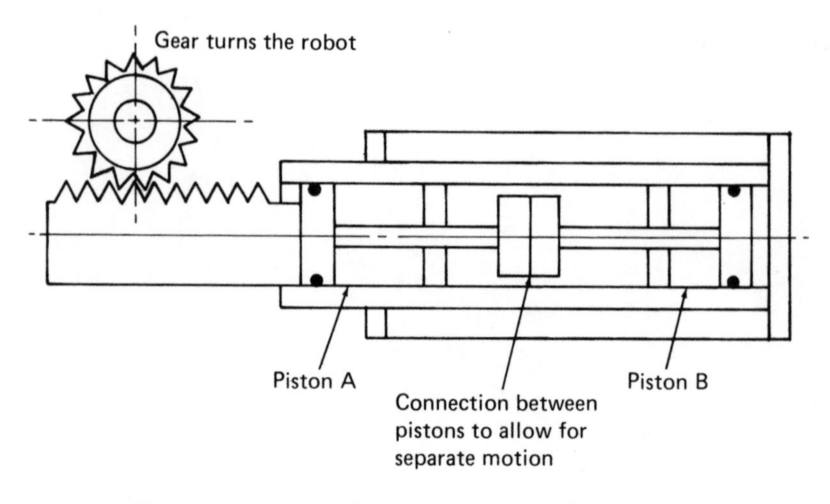


**Figure 5.3.** A hybrid device for rotation of a robot base.

## SIMPLE OR COMPLEX

While any pick-and-place type robot with multiple stop positions might be considered a robot hybrid, we should establish some criteria for separating the most simple of the units from the more sophisticated. In today's market, there seems to be a natural dividing line between the pick-and-place robots with only one or two additional motions, and robots of the more powerful type.

Figure 5.4 shows a robot with multiple arms currently being manufactured. All of the arms can be commanded with the same signal (or, if the operator desires, with different signals), and all of the arms travel on a common rotation and lift base. To the robot designer, these arms can all be treated as a single unit, while the programmer may make use of the extra ability of the arms to reach a larger combination of positions.

A typical pick-and-place robot with three axes of freedom (left-right, up-down, in-out), would have exactly eight positions in its reach. The formula to calculate the number of unique stopping points is $A_1 \times A_2 \times A_3$, etc., where A is the number of stop positions on the first axis, $A_2$ is the number of stop positions on the second axis, and so on.

If we treat the three-armed robot as a three-axis machine, we would calculate the number of stop positions as: $2 \times 2 \times (3 \times 3)$, the base having two stop positions, the lift having two stop positions, and each of the three arms having three stop positions. This robot has in its reach 36 unique positions. For our purposes, we will consider that a robot is the

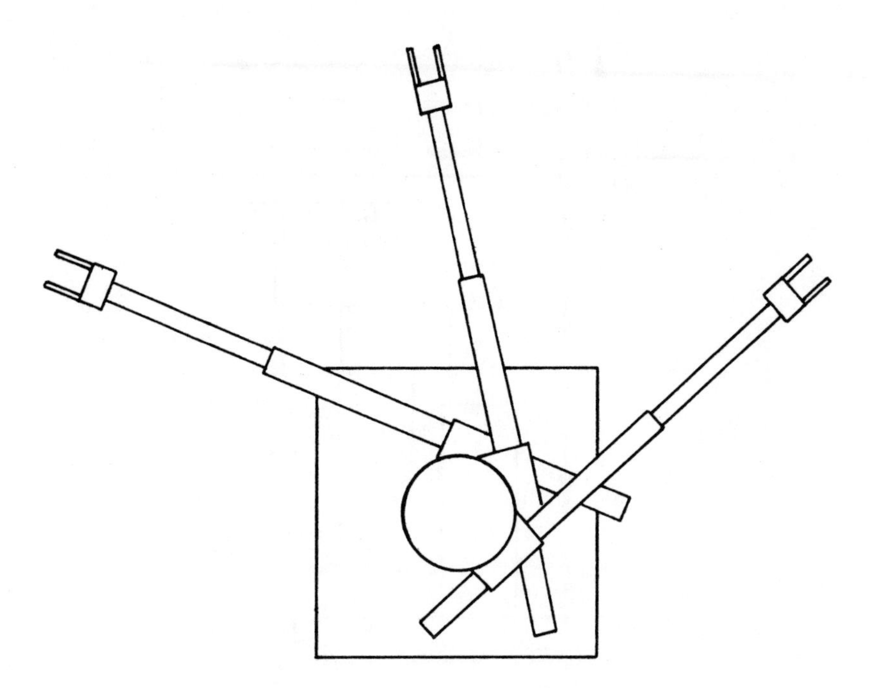

**Figure 5.4.** A three-arm hybrid robot.

hybrid type if it can reach as many points as can be expressed by the formula $3^A$, where A is the number of axes.

We see, however, that a robot specifically designed to be a hybrid might have many more stop points than is required by our definition. One example is the robot in Figure 5.5, which has 2,304 unique positions, as it is currently being built (six lift, twelve rotation, eight extension, two twist, and two slide motions).

The formula for the number of stops can be applied to the servo-type robot. It is not uncommon to have as many as 5,000 unique positions for each axis. A 5-axis servo-controlled robot can reach ($5{,}000^5$) 3,125,000,000,000,000,000 unique positions. There is sufficient difference between the amounts for the servo-controlled robot to be considered a separate class.

There is more to the difference between the robot hybrid and the servo-controlled robot than simply the number of positions they can reach. Like the simple pick-and-place robot, the motions of the hybrid from one point to another are not controlled. When the robot controller is commanding the axis to move, it is simply turning on the power and

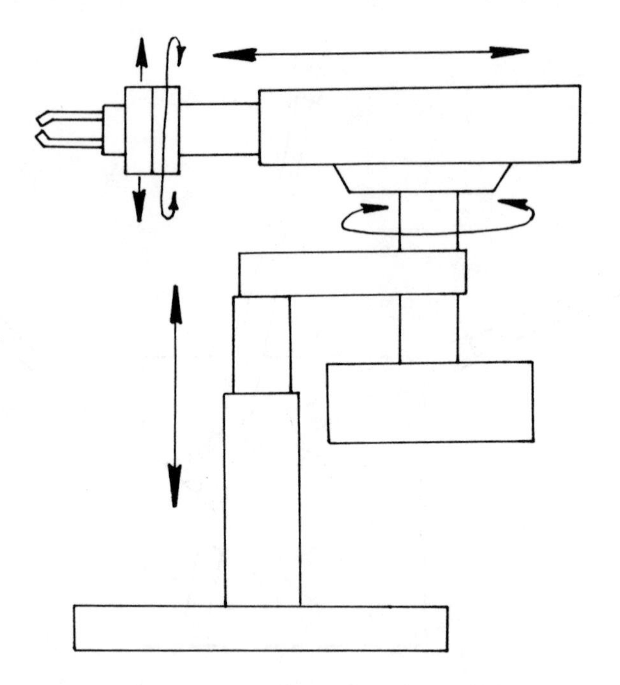

**Figure 5.5.** A hybrid robot having 2,304 positions.

then waiting for the action to finish. The path the hybrid will take in its journey to the new position is not controllable. The speed of the arms on a hybrid also is not controllable and must be adjusted manually. The servo-controlled robot has a unique feature in its ability not only to reach a large number of points, but also to control its path and speed in getting there.

## MODULAR ROBOTS

Some manufacturers make their robot components in modular sections. It is possible to buy the desired sections and assemble them to form a unique combination that exactly suits the buyer's purpose. This modular arrangement allows the robot manufacturer considerable flexibility in the selection of sizes and types of motion offered to the customer. It is possible to make the base, upon which the other components will rest, very heavy duty for carrying many components or light and fast acting if only a few components are to be added. By properly designing the flanges which will join the sections together, it is possible to join any subsection to any other subsection. Figure 5.6 shows a base with multiple components available for addition.

It is also sometimes advisable to assemble a robot with more than one type of power component. For example, a particular robot may need a very forceful lifting section but need only a small amount of force for indexing back and forth. At times, it may be economical to provide the lifting portion of the robot with hydraulic components and the shuttle portion with air components.

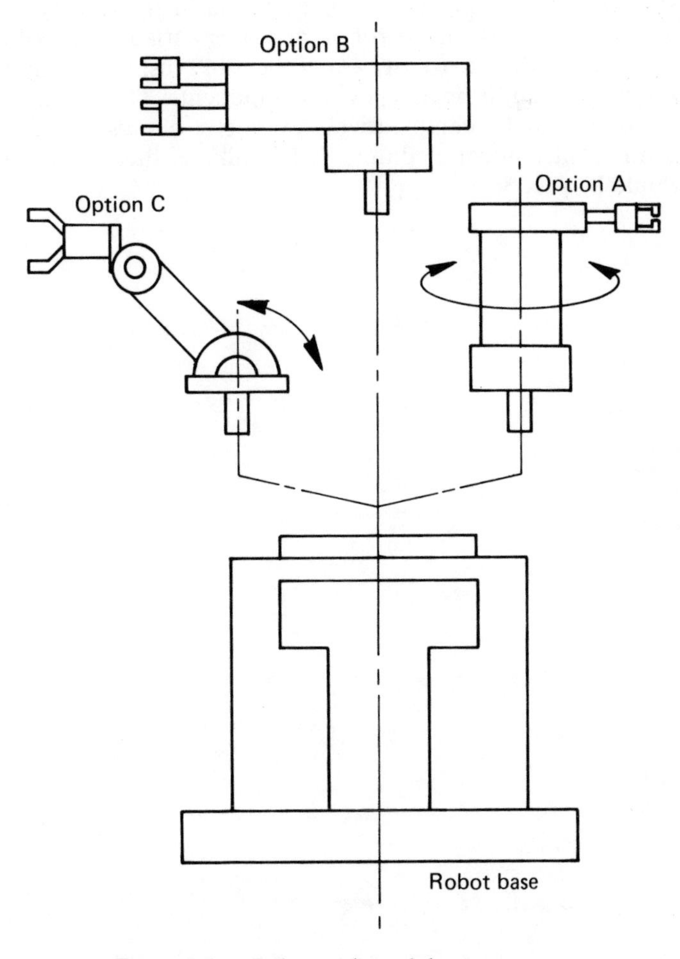


**Figure 5.6.** Robot with modular components.

The same may be true with electrical functions and air functions, or any other combinations. Indeed, it is common for electric robots to still have pneumatic gripper fingers. In some instances where a hydraulic level of force is required, it is possible to use an air-powered robot. Rather than using a pump and tank for producing the needed hydraulic pressure, a special air and oil pressurizing chamber is used. A comparatively large volume of air is used to compress oil in this chamber. The smaller amount of oil compressed is at a much larger pressure than the air being supplied.

Because of the volume differential, there is no creation of energy, merely a transformation. It is possible using this arrangement to convert low-pressure air, such as already might be found in the factory, to high-pressure hydraulics for use in the robot, at a substantially reduced cost.

It is conceivable that a robot would be built using air, hydraulic, electrical and other additional types of components. The control system for this robot would be comparatively expensive because of the use of the many different control mediums. It is unlikely, therefore, that such a robot would be practical.

# CHAPTER 6

# ROBOTS: FACT AND FICTION

Because there has been so much written and shown on the motion picture screen about robots, it is sometimes hard for the general public to distinguish between the true facts of modern-day robots and the myths. The following is some simple basic information about robots (some of which may not be considered as fact, since it is the estimation of the author):

## BASIC ROBOT INFORMATION

- Prices range from $2,000 to $250,000.
- Sizes range from a ten-pound robot with one-pound payload and six-inch reach to a 6,000-pound robot, with 2,000-pound payload and 20-foot reach.
- Robots are mounted on track systems enabling them to move about.
- Robots are assembled from components used elsewhere in industry.
- Some robots use computers for control, some do not.
- Robots may be hydraulic, pneumatic or electric.
- There were about 30,000 robots in the world in 1982.
- There were about 6,000 robots in the United States in 1982.
- Robotics is one of the fastest-growing industries in the world.
- There are over 200 robot manufacturing companies.
- There may be as many as 350,000 robots in the United States by 1990.

- Estimates of the number of U.S. workers displaced by robots by 1990 range from 1,000,000 to 15,000,000.
- Estimates of the number of jobs created by robots by 1990 range from 250,000 to 2,000,000.
- The two countries using the most robots per worker population (Japan and Sweden) have virtually no unemployment and export a large percentage of their products.

The robots being used today perform a variety of tasks. The following list indicates the relative frequency with which each of these jobs is accomplished, ranging from most common to least common task.

- Materials handling
- Machine loading
- Palletizing
- Spot welding
- Arc welding
- Paint spray application
- Deburring of parts
- Grinding
- Glue application
- Cutter guiding
- Inspection
- Assembly

A practical and real robot task is given in Chapter 16; and, although the example is a composite of several real robot installations, it illustrates the capability of a typical modern industrial robot.

## IMAGINATION STAGE

In some news media, one can find pictures of very dramatic robot developments. There are, particularly in magazines dedicated to robot technology, articles and pictures showing robot hands, similar to the human hand, with articulated grip. Robots that mix cocktails. Robots that retrieve beer from a remote refrigerator and open it for the user. The pictures notwithstanding, these robots should still be considered as fiction.

To say that we have a real functional robot is to equate our product with any of the other machines used in our society. Automobiles are real–not just because one of them has been built and they are therefore theoretically possible to produce–but because an automobile may be produced in quantity and purchased by others. It may be strictly accurate to say a college professor's development of a robot with new and exciting features constitutes its entry into reality; but, as a practical matter, such a development often involves an abundant use of unusual components and technology.

Although a robot may possess a vision system sophisticated enough to locate the refrigerator, find some beer, remove its target, and give it to the programmer, how much effort and expense has been expended to develop this process? Is this a practical development, or is it something we have one time been able to produce but should not expect to consider part of our daily lives?

We might say it is possible for us to vacation on the moon, as a few people have gone there and returned safely. Or, it is possible for the United States to completely cease its use of petroleum products and convert to solar power. Can either of these possibilities be considered a reality in the near future? When one considers the long and complicated process of robot research and development, one should not take the musings of a robot developer as the current state of the art.

We do not wish to discourage anyone from attempting to improve on current technology, nor to limit creative thought about the future of robotics. However, laymen should be conservative in their expectations of the science and benefits of robots. Let us consider some of the misconceptions of the robot in its practical use.

## ROBOT MOBILITY

One characteristic that distinguishes between a science-fiction robot and a real-world robot is its ability to walk and move about. In almost all instances in the modern use of robots, the parts the robot is to manipulate or the work the robot is to do is brought to the location of the robot. The robot has a limited ability to change its position. Sometimes robots are mounted on tracks, sometimes they move across the floor, but their ability to move is limited. The robot is restricted to those areas where it will carry out its normal tasks.

A typical science-fiction robot is quite animated. It has legs, arms, and eyes to see where it is going. The only restriction placed on it is that it must follow some preconceived plan of duties. At the current state of the art, it is not feasible for practical robots to have the mobile freedom of their science-fiction counterparts. It is too easy for a malfunction to occur. An then we would have a berserk robot roaming the streets to be hunted down by the police.

With a limited ability to move (for example, to move only within the bounds of the manufacturing plant), it is conceivable that robots will be given legs, camera eyes to see where they are going, and the microprocessor thought process necessary to move themselves about. After much research and development, some robots of the future might incorporate this type of design, but, the robot would not be restricted to its own small working place; it would have a large range in which to work.

Examples would be robots that pick crops, work in the forest, or attend to hazardous duties such as fighting fires.

## A ROBOT IN THE KITCHEN?

One of the most commonly desired robots by the general public is a domestic servant. When one robot company president gave a speech at a public gathering, he was asked when his company would develop such a robot. The audience enthusiastically applauded, indicating such a device would be well received, indeed. Further, the questioner indicated a willingness to pay up to $5,000 for such a device. In terms of domestic home economy, a home robot should not be more expensive than other domestic equipment. One would expect to pay no more for a robot than for an automobile or for the remodeling of a room.

However, a domestic robot would be extremely complicated in its design. A robot to wash clothes, wash the dishes, run the vacuum cleaner, clean the bathroom, take out the garbage and let out the dog would need a great number of abilities.

It would need the ability to pick up and manipulate fragile objects, to locate these objects placed randomly around its working area, and to differentiate between objects (the difference between a rug to be washed and the sleeping family dog, for instance). This robot would need to use other equipment, turn on water taps, plug in the vacuum cleaner, measure soap powder, turn on the washing machine, open doors, tie plastic bags. All of this is in the realm of possibility for modern technology. It is also conceivable that a robot would possess sufficient memory to be programmed for each of these individual responsibilities.

If there were some input mechanisms or timing functions that could tell the robot when to begin these tasks, we would have a robot with the ability to be a domestic servant. There are some other important considerations, however. If the robot's timer indicated it was time to wash the dishes, the occupants of the robotized household would need to exercise great care that they were finished eating. If the bathroom were being cleaned, care would have to be taken that important papers or money had not accidently been knocked to the floor.

Sufficient safeguards could be programmed into a robot using existing technology. So, it is not the tasks themselves and their technological answers that limit our ability to create a domestic robot.

The biggest constraint is price. The would-be customer felt a domestic robot should cost up to $5,000. However, examining current industrial robots of comparable complexity, we see that our domestic robot would cost much more than $5,000. A dishwashing robot (Figure 6.1), for example, that would be able to move over a countertop, locate plates placed randomly on it, pick up and manipulate these plates, put them into a dishwasher or wash them directly, and then place the clean plates in a cupboard would approximate the sophistication and ability level of a robot costing $95,000.

It is possible to build our domestic robot, but to have all the features would require great expense. The research and development costs for this robot construction might be $10,000,000, and the cost of the physical components together with the assembly of the robot might

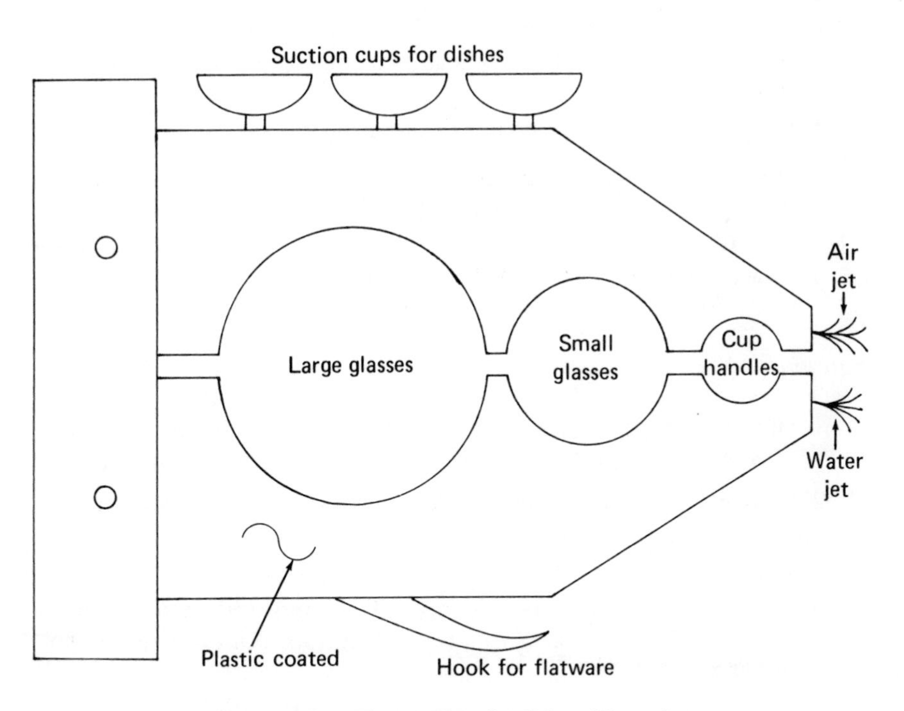


**Figure 6.1.** The gripper of a dishwashing robot.

be $200,000. Assume that to break even—and not allowing for interest paid on our investment money—we must sell 100 robots at $300,000 per unit. How many households would purchase a domestic robot for $300,000?

This $300,000 placed in a bank at some high-yield interest would provide a sufficient return to hire a full-time maid. Most families, given the option of having a flesh-and-blood maid or a cumbersome and still limited robot—forcing them to eat only at certain times and putting great fear and frustration in the mind of the household dog—would choose the former.

Consider the following feasibility study for the dishwashing robot.

## FEASIBILITY STUDY

### Robot-Loaded Dishwasher

*Job Description*

1. Pick up randomly placed dishes from counter, place food-side down in dishwasher, place utensils in center basket.
2. When washer is full, stop loading, add soap, close door, push start button.
3. Ready for other jobs while dishwasher cycles.
4. Open door, open cupboard, remove dishes and place in correct position in cupboards; when out of dishes close cupboards, return to 1.
5. If washer is not full and dirty dishes are all placed: close door, ready for other work.

*Equipment Needed*

1. One-arm, robot. Axes for motion: up-down, in-out, left-right, twist end of wrist, bend end of wrist. All axes servo controlled.
2. Gripper able to grasp and hold: plates (four sizes), glasses, bowls, pots, pans, pizza tins, pie plates, cups, stemware, various serving utensils, knives, forks, spoons.
3. Vision package able to recognize all of above elements in space and able to direct arm to proper location. (Option: monochromatic spot detector.)
4. Track system to remove robot to out-of-the-way position when not in use.

*Computer Memory*

1. Sixty-three different shapes.
2. Two dishwasher baskets configured as 22 X 15 positions + 12 positions in center tray.
3. 117 positions in cupboard, up to 12 layers.
4. Plastic tray in silverware drawer with 9 positions.
5. Option: special memory to facilitate most dense packing of dishes.

*Price*

| | |
|---|---|
| Robot | $50,000 |
| Gripper | 10,000 |
| Vision | 30,000 |
| Track system | 5,000 |
| | $95,000 |

## A LAWN-MOWING ROBOT?

The robot type second most requested by the general public is a variation of the domestic robot. Indeed, it could be a domestic robot with additional tasks assigned to it. It is the dream of many a would-be Saturday afternoon golfer: a lawn-mowing robot. Consider the lawn mowing function. Because the surface area of the lawn is so great, it is necessary to have a rather long, sharp blade to do the cutting in a reasonable length of time. On a power mower, the blade is turned at very high speed. This blade would be potentially a lethal weapon on a runaway domestic robot.

Because robots are so obedient in their programming, it is easily possible that a robot, while mowing the lawn, would take its command too literally and cause great destruction to people and property. By some mistake in programming, the robot may well decide that a particular place to be mowed happens to be in the bounds of a neighbor's rose garden. Worse, if the neighbor is working in the rose garden, we have the makings of a very large lawsuit. Although the task of mowing a lawn fits our domestic model of a good installation for a robot, it is one of the tasks that will remain a human function for a long time to come.

# CHAPTER 7

# A COMPUTER FOR A BRAIN

A robot needs an equivalent to human muscle to do its work. It also needs the equivalent to a human brain to control the muscle. For a robot, the brain is a computer. The robots we have been discussing are of various levels of complexity. Likewise, the computers controlling them are simple or complex, as needed.

## SEQUENCE CONTROLLERS

The simplest type of computer used to control a robot is similar to music box chimes and is called a contact drum. It is a rotating drum, having several bumps which engage electrical contacts (Figure 7.1). The drum is driven either by a motor, much like a clock motor keeping continuous speed, or by some type of stepper motor that does one step, pauses a certain length of time, and then does the next step.

A contact drum can be used to control something like a washing machine. Here, various valves open to allow water into the chamber, to drain water out, to dispense soap, to drive the agitator driven back and forth, and to complete cycles as needed. The drum rotates and causes each control for the separate functions to be turned on or off at the correct time. In a very simple robot, this type of control can be used to move the arms of the robot, to turn the various other functions of the robot on or off, and to initiate and stop the sequence.

A slightly more complex form of controller, but nevertheless called a sequence controller, can be reprogrammed to do different functions, each following the one before it. The sequence selects the predetermined motions and sets their order. We have no control over the motions themselves once the selection is accomplished either by some drum arrangement, as previously mentioned, or by some electronic function. Each step of the sequence can initiate a number of events. If a drum is used for a sequence controller, the actual switches that engage the electrical

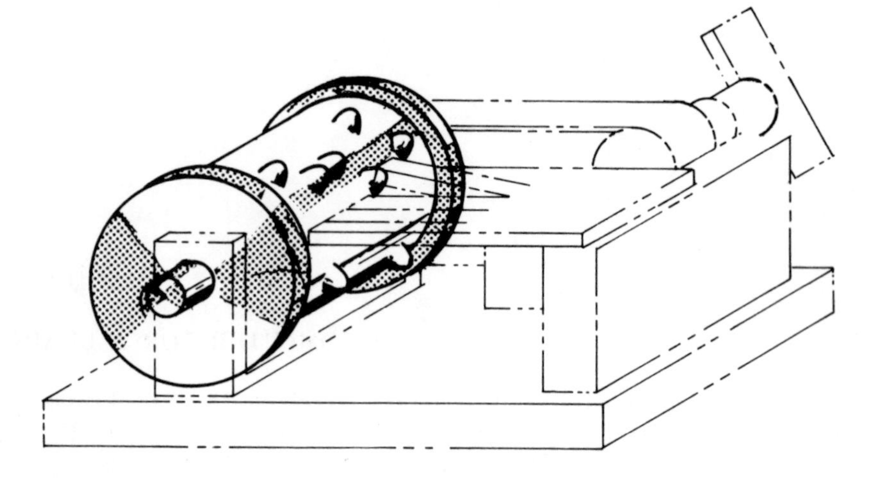

Like a music box drum, the contact drum rotates and makes various connections. In place of the music bars are valves or electrical contacts.

**Figure 7.1.** A rotating drum music box.

contacts are changeable. One such mechanism has small pegs that are placed in the drum. As the drum rotates, these pegs come in contact with various switches. By changing the position of the pegs, we program the robot in a rudimentary form. This type of programming is similar to machine-language programming in computers. Since the programmer is making the mental effort to create the individual sequences, there is no need for a computer-type language.

Sometimes the computer is not electric. When a pneumatic robot is to be used for a simple task, a pneumatic controller can be used as the computer element. *Air logic elements* are small valves that can be arranged in much the same way as electrical transistors to provide control functions. It is often easier to use this type of pneumatic logic to control the robot than to use electrical control signals that must be converted to air signals before being used by the robot. If the signal to start the robot is also an air signal, using air is even simpler.

Figure 7.2 shows how an air logic circuit can be used to control a robot arm without the use of intermediary equipment. The signal coming into the robot is from the machine it is unloading. As this signal is altered by the pneumatic logic, it causes first one arm, then another, to move to pick up the part. Several of these circuits together can control the entire operation of the robot. Programming of this type of

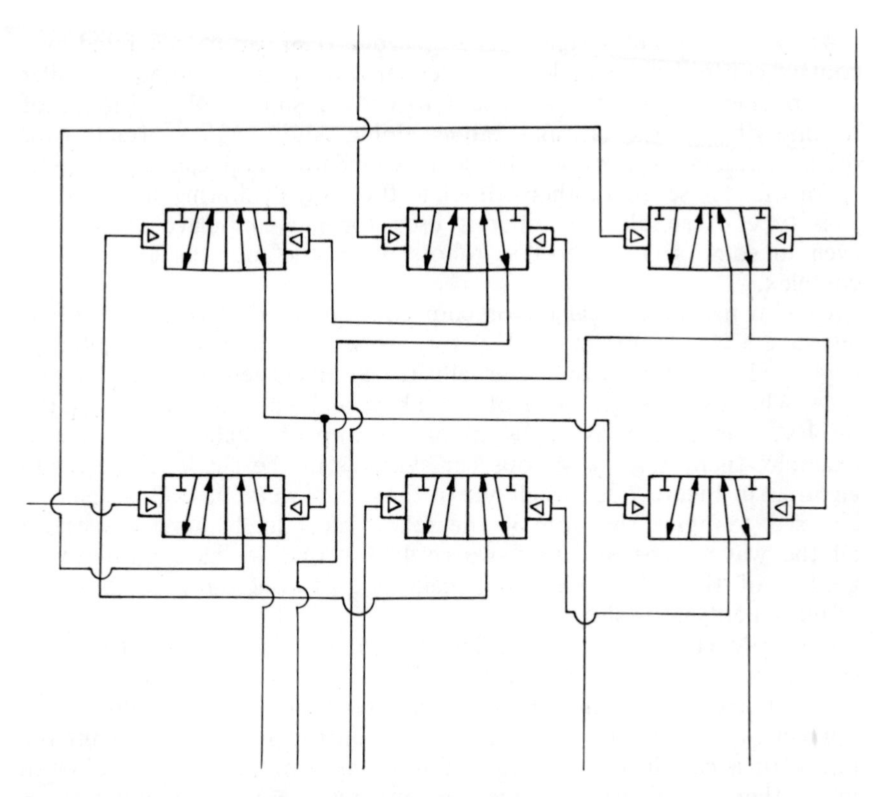

Small air valves can perform the same logic functions as a computer. If a robot is operated pneumatically, it is sometimes simpler to have the control work directly off air rather than to convert an electronic signal each time.

**Figure 7.2.** An air logic control segment.

controller is usually done by changing the connections of the air lines. This is similar to the type of programming done on very early electronic computers.

A major drawback to this type of control is the limitations of the air logic elements themselves. It is difficult to make a circuit of any great complexity using air logic. The units function at a speed over a thousand times slower than their electronic counterparts. An air signal propagating through even a few logic elements must be amplified before it can be used. These factors limit the use of air logic to only the simplest of robots and to the most basic jobs.

When more complex functions are needed than can be generated by a contact controller, a simple computer must be used. A computer similar to a pocket calculator is sufficient to drive a simple robot. Instead of manipulating mathematical symbols, doing addition or subtraction or other mathematical work, the small calculator chip can turn inputs on or off. Those inputs then stimulate the individual arms of the robot. This type of simple controller is used for pick-and-place robots and even in some servo-controlled robots when the motions are not too complex.

To visualize how a calculator chip could be used to control a robot, lets take a closer look at a calculator. We see that there are a number of keys used to run the display, usually the numbers zero and one through nine. When we push any one of these keys and a number appears on the display, the number is made up of a group of smaller symbols. For example, there may be various bar symbols in the shape of the figure eight, as in Figure 7.3. If each one of those bars were, instead, a signal to be used to move the arm of the robot—most likely moving the arm all the way in one direction—we could say that we have push-button control of the robot. We could make it go through any combination of movements we wished.

The calculator chip controlling the robot also can do functions similar to addition and subtraction. If we want more than one arm to move at a time, we can initiate a computer function that is almost the same as saying one plus seven. The robot controller would first make the one with a certain combination of motions and then make the seven in another combination of motions to create a new combination of motions that is the sum of the other two. In the case of the robot, the number eight would not be the sum of one and seven. Rather, the motions would be added together, and the sum of one and seven would be an inverted letter U.

Once we know that this motion is what we want, we can make one and seven a single line of our program. For purposes of controlling a robot, it is necessary only that the calculator do one complete function, know it is finished, and then continue on with the next function. It is actually simpler work than is done by a pocket calculator.

The chip must have some control feature that will allow it to sequence from one operation to the next, because a continuous record must be kept of all the additions and subtractions. Since the number of steps we wish a robot to do automatically can be quite high, several hundred, for example, it is necessary for a robot controller to have more memory than most calculators.

The memory controlling the robot is used to store the unique set of instructions for a given program. This is different from the way a pocket calculator is used. In a calculator, once we have added the numbers one, three, and five, and get the total nine, it is usually not necessary to

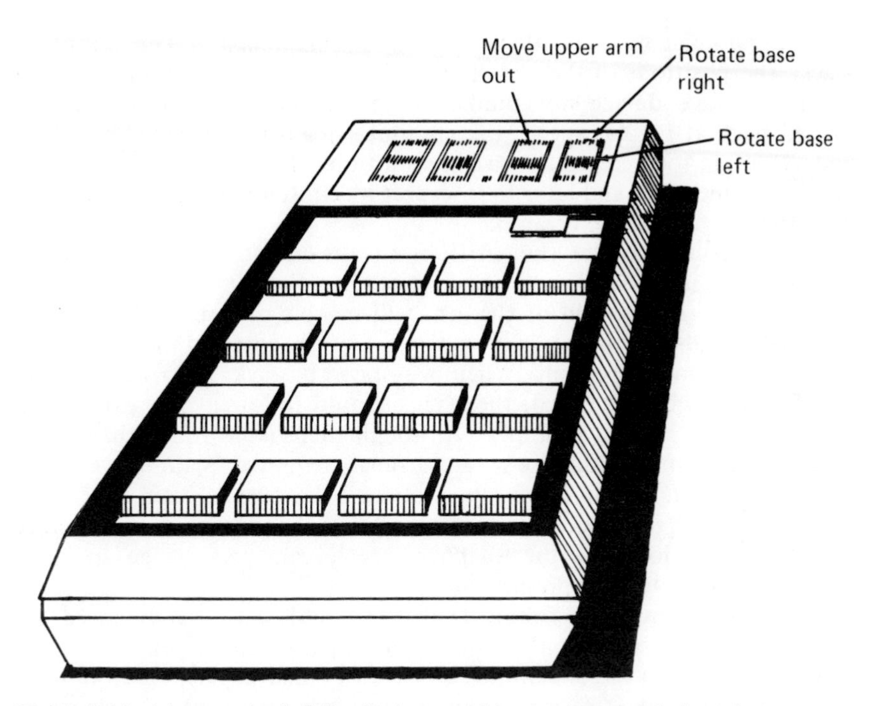


The individual bar on the L.E.D. display could be used to signal the robot to move one of its arms. The same sequence of numbers would be used over and over as a program.

**Figure 7.3.** A calculator example of a simple electronic control.

use that same sequence over and over again. Perhaps in the future, if we have forgotten the total, we will add that same sequence of numbers again—but only if we have forgotten.

With a robot controller, we may wish a sequence of events we conceptualize as "event one," "event three," and "event five" done over and over again, hundreds of times. So, it is necessary for the calculator chip controlling the robot to remember entire sequences, rather than simple instructions. Therefore, a calculator chip controlling a robot will be more complicated than one controlling a calculator.

## TRACKING CONTROLLERS

The simplest robot computer used to control a servo-motor type robot is a tracking computer. The robots using this type of control are programmed by manually moving the arms in space. The computer tracks the positions of the arms in real time and stores the positions on a moving storage device not unlike a tape recorder. When the tape is played back at regular speed, the computer reads the recorded positions and attempts to recreate the original motions. The computer is used to create the memory and later to interpret its contents. It has no ability to alter the program, once created.

We can conceptualize the creation of the memory by the example shown in Figure 7.4. We see one axis of the robot free to move in and out. A stylus at the end of the arm will make a groove in the rotating disk. While the disk is moving, we may grasp the arm and move it in and out, creating our program. When we start the disk again at the *start* point at the previous speed, the stylus should follow in the groove and reproduce our original motions. (Although there is no robot that works in this way, some robots use a cam arrangement that is similar. Each of the arms of the robot can be positioned by a turning cam, the cams all being driven by a common motor. The robot designer will dictate the arms positions by the shape of each cam. A change in cams allows the robot to be reprogrammed.)

In a servo-controlled robot, the stylus and disk are replaced by electronic memory, but the robot still tracks the original motions. The motions themselves cannot be altered or rearranged unless another program is created. This type of programming is adequate for operations such as spray painting or gluing, where the accuracy of the arms is not

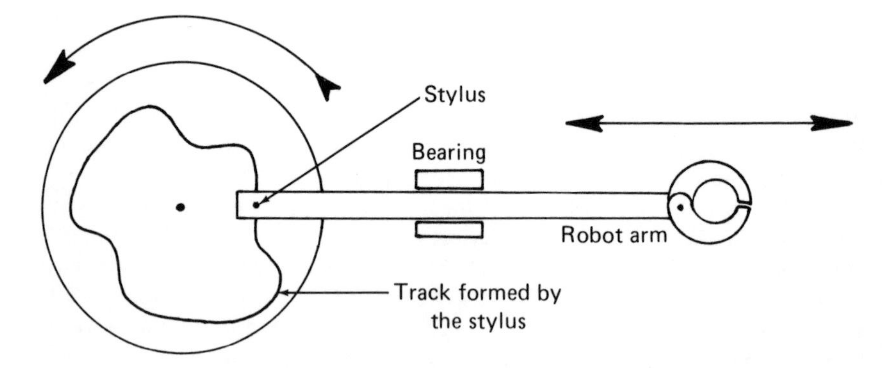


As we manually move our tracking model robot, a stylus makes a groove in the rotating disk. When the disk is "played back," the arm will repeat our original moves. In a real tracking-type robot, the storage of the movements is electronic.

**Figure 7.4.** A model of a tracking-type control system.

great and the exact speed with which a task is accomplished is not important to its ultimate success. The limitations of this style robot, however, are that a human being must at least once accomplish the task manually with the robot for the program to be created.

## MORE ADVANCED COMPUTERS

If the work to be done by the robot is more complex (and usually if the robot is the type that is servo controlled), a more sophisticated computer than the tracking type is required. The computers used on very advanced robots are just as sophisticated as any computer used in industry or science. The registration- and informational-programming types use very similar computers, their main difference being in the method of programming. Many of the improvements made in semiconductor memory and in very large scale integration are currently being used to control robots.

It is also possible that very large-scale computers, such as the main computer for a large company, can be used to communicate directly with the computers used as part of the robot. Hundreds of robots can be controlled by a single master computer giving instructions to each of the robots. The individual robot still keeps its own computer, however, as the information processed by a moving robot is too rapid for a separate computer not dedicated specifically to that task.

The robot's dedicated computer uses a great deal of memory in comparison to the simpler type of computer. It is also very sophisticated in its ability to process information from a multitude of sources and to use that information for decision making. These computers are also more sophisticated in their use of instructions, the way they carry out their instructions, and the number of instructions they are capable of accomplishing in a unit of time.

A sophisticated-type computer controller is capable of performing thousands of instructions per second. A servo-controlled robot moving at high speed may have its motors feeding back information into the computer at these rates, and the misplacement of one bit may make the robot stop in the wrong position. If the computer loses track of the robot or is slow in processing information, the robot may physically destroy itself or harm its work place. For this reason, the computers must be very fast.

The current state of the art of computer science allows the computer to process information at extremely high speeds. Currently, these speeds are greater than needed by an individual set of robot arms. The computer portion of the robot is now more advanced. In the future, however, when robots are doing increasingly complex functions, it may be that computers will need even more power to accomplish their task.

The type of information being processed in a sophisticated computer is, again, the same type as is used for mathematical manipulation. A series of inputs is given to the computer; and, for the given robot, this is a set of instructions for accomplishing each task. The computer, because it is of a type used for other purposes in industry, manipulates this information as it would any number.

A given piece of memory for a computer might be only a series of zeros and ones. If we were to examine a particular piece when this information is processed, we would see how it becomes a series of motions carried out by the robot.

The transformation from electronic state to actual robot motions usually is carried out in several steps. The processed information is routed to another card (circuit board) where it is translated into signals understood by the individual components. If it is a servo-controlled robot and the command on this particular piece of information is to move motor number three, then the translated command may take the form of a voltage on one of the motor's controls. In a pick-and-place robot, the information may be the command for a valve to open, causing one of the arms to move. This translation may take the form of a pulse to one of the flip-flop elements controlling the valves.

In a very sophisticated robot, the number of translations between a command of the chip and the actual accomplishment can be quite high. Each component in the string carries out its particular function. Some read the signal and route it to the particular part of the robot that is being controlled at that time, and some of the components temporarily hold the signal until the other components are able to accept it.

The general categories of the additional boards may be summed up as: memory, computation, communication.

## DIFFERENT FUNCTIONS OF THE COMPUTER

### Memory

The single greatest memory capacity of the robot usually is taken up in the basic instructions set; that is, the instructions the robot computer will use to carry out the physical manipulation of the robot. When the robot is created, its computer is given a memory component capable of keeping the master set of instructions over a long period of time. This master set of instructions, often called the *executive memory* or *executive program*, always must remain in the computer. If it is erased or somehow lost, the computer will not be able to control the robot.

We must differentiate between this type of memory and the memory the user will create when a program is made. The executive program is usually written only once and is the same for all robots of that particular type.

We can compare robots based on the strength and size of their executive memory. For a very simple robot, a control system need only turn on and off valves or similar single-state devices. The executive program may be as simple as a rotating drum. The executive program for a large sophisticated servo-controlled robot may well have thousands of lines of computer instructions and may do hundreds of logical steps for each instruction.

In a registration-type robot, the executive program must contain instructions necessary to record the positions and to alter the sequence of the program. In an informational-type robot, the executive program must not only process the recorded points and alter them, but it must also be able to receive the purely mathematical information of the programmer and to configure programs from it. The distinction between the two types can be almost entirely based on the power of the executive program. Two robots might have the same microprocessor and the same set of physical arms; yet, if one of them has the ability to interpret the input of mathematical data while the other does not, then the robots can be considered different types.

To examine the difference between the two types of robot controls, let's consider a delivery truck. The truck itself symbolizes the physical aspects of the robot. The control functions are taken up by the driver, with the executive program being the driver's ability to navigate on his own.

We have a certain type of driver who can take us any place the truck will go, but we must go with him the first time. Once he has gone to a particular place, he writes the directions to get there in a special book. We need now tell him only the name of the place, and he can return there by himself.

A program for this driver consists of a list of instructions giving places and the order in which they must be visited. We can make a new program by rearranging the places or by going to new places to show him the way. We may be able to give the driver some additional tasks to do on his journey. We may ask him to wait at the bakery for a special cake that is being prepared. We might want him to wait at the cleaners until three o'clock and then go to the butcher's. This driver can understand our instructions to rearrange the route and to do the other tasks, even if we are not present. He cannot follow our instructions to go to a location unless he has it recorded in his book. (An exception to this is that the driver always can find his own garage without our help.)

There is a different type of driver we can call an informed driver. He has all of the ability of the first driver, plus the ability to find places on his own. We can give him information on a destination, and he can go there the first time himself. The information may be of a special type, such as latitude and longitude, but he can get there. The two drivers may have the same intelligence, and the trucks may be identical, but the ease with which we can direct them makes for a difference in usefulness.

The two types of drivers are like two types of robot controls. A registration robot must be manually driven through each program step for the program to be created. An informational robot can be commanded by direct entry of the information.

To extend our analogy to the other two types of robot controls, we can imagine that a tracking control robot is a truck that will only follow the exact route we give it. The driver does not make any notes for himself, and he can only duplicate what we showed him on a previous trip.

A sequence controller is like a truck traveling on railroad tracks. We can change the order of the stops but have no control over the exact positions of the stops. A contact drum controller is like a truck on tracks, with the tracks going around in a circle.

While a computer executive program is stored in some memory device, a program itself is software.

The executive program can be stored on a device such as an EPROM chip which allows a master program, once written, to be duplicated on hundreds of chips at very low cost. It may be put on some type of semiconductor memory specifically created for this one task. This type of chip is very expensive, however, and is not being used as much today as in the past.

Another memory function of the computer is what is called the *read and write memory*. It is this portion of memory that is established by the user when a program is made. It is usually provided blank to the customer, ready for use. In some robots, this memory remains while the power is shut off. In other robots, the read and write memory will only be retained if the power is on continuously. If the robot is to be turned off, the memory must be put on some other device, such as cassette tape or floppy disk.

The type of memory element used is sometimes called *random axis memory* or *RAM*. Each time a different program is written for the robot, a different set of instructions is placed on the RAM memory. These instructions may be entered as a high-order language similar to BASIC or FORTRAN, in which human-type words are used to describe the motions of the robot. It is possible in such a system to make very few key strokes and yet enter a very complex program.

In some other systems, it is necessary to use programming similar to machine-language code, although it is entered through a computer-type console.

Regardless of the type of system used to enter the memory, the read/write memory can be changed for different programs. In some robots, memory must be created before any motion of the robot is possible. In these off-line programming robots, the robot's motion does not respond directly to the push of buttons by the operator. Instead, the memory is created by the push buttons, and the memory causes motion in the robot.

In some other robots, which we classified as registration and are sometimes called *teach-type* robots, the robot is manually positioned by the use of jog buttons. Then, the key pad is used. This position is registered and numbered for a particular instruction. Here, it is not possible to enter a program using only a mathematical structure of positions, such as an X-Y-Z coordinate axis system. Rather, the robot responds to the computer by giving its position, and it is this information which is recorded in the memory rather than some artificially defined matrix.

Another function of robot memory, other than storing the executive program and storing programs written by the user, is the temporary storage of information used in processing the motions of the robot. At any given instant, the robot computer is examining only one piece of information. Since there is always more than one motion produced by a robot, there often will be multiple input signals from other equipment. Some information must temporarily be stored until the computer is ready to receive and process it. This type of memory sometimes need last only a few thousandths of a second. But, because some signals are so short-lived, it is necessary to have a memory function for this short storage of information.

## Computation

Another function of the computer is the computation of positions of the robot and other information needed to produce motion. If a particular robot arm is to move from Point A to Point B, a certain combination of motions in the motors is necessary. This can be very direct for the computer to calculate. If the path from A to B is to be by some nondirect motion, some interpolation might be necessary.

Ever increasingly, the demands put on the motions of the robot are artificial in terms of the natural movements of the robot. A particular robot may have a base that rotates about a center point, but the motion that is desired might be a straight line across the natural arc of the base. To accomplish the straight line, some in and out arm must compensate for the arc of the base. Each position along the line requires some mathematical manipulation of two components (Figure 7.5). If the line is to be extremely straight, the number of computations can be very large. For instance, to move six inches along a straight line with an accuracy of 0.005 inches, a particular robot is required to execute 2,400 instructions. Each of these instructions involves a geometric calculation involving arc and length.

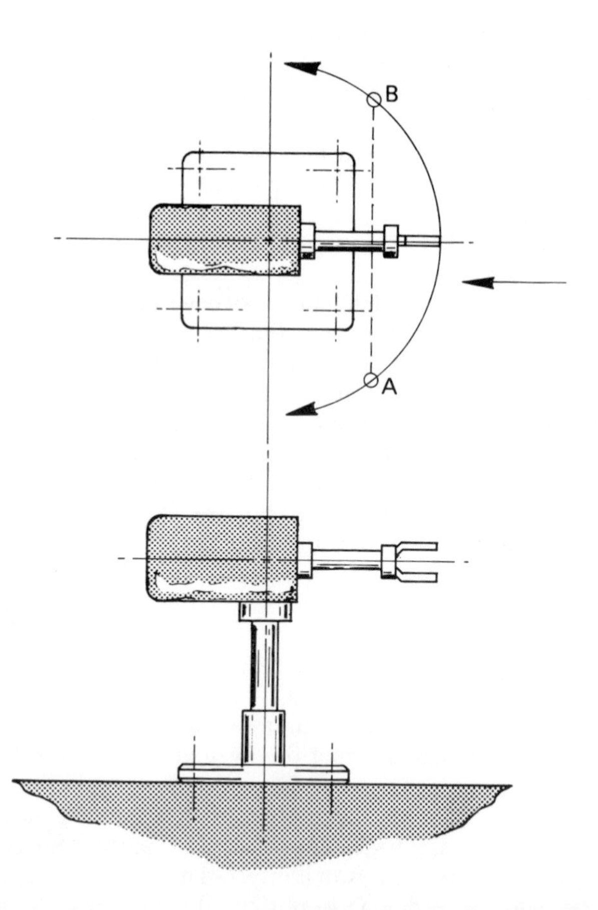


The natural path of the robot from A-B is shown by the arrows. The desired path is shown by the broken line. A large number of robot calculations might be necessary to form the desired motion.

**Figure 7.5.** A robot using linear interpolation.

In addition to computations needed for position, there are computations needed to control the speed of the robot. Most motions of a sophisticated robot are not direct in their speed. Each of the axes must be accelerated in a predetermined way. When the robot nears its required position, each of the axes must be decelerated. This is especially necessary when the robot is handling extremely large or fragile loads. The computation of the exact speed for each of the axis, the RPM if a motor is used, or the exact flow rate if some linear displacement element is used, can be very demanding if done in real time. If the motion takes only one second, but during that second there must be an acceleration, a travel at a constant speed, and a deceleration, the calculations must be done very, very rapidly.

Another function of robot computation is decision making. Most modern computers have the ability to receive inputs from the outside world and alter their programs based on the state of these inputs. It is possible to have an input A and an input B. The robot can do a particular task if it receives input A, a different task with input B, and still a third task with both input A and input B. The computers are capable of the usual logic formations *and*, *or*, *not*, and *nor*, and these can be used by the programmer to accomplish very sophisticated tasks.

This decision-making capability, which in computer language would be similar to the conditional jump routine, is also necessary when the robot is operating with other machinery. Unprogrammed circumstances may arise with this machinery where the machine may jam, it may run out of parts, and it may begin to produce parts at a rate incompatible with the robot's ability to handle them. When these decision-making factors are properly programmed into the robot, there is no need to have an operator on hand at all times. The robot can do its own decision making.

The final factor we will consider in the computation done by robots is the interpretation of sensory inputs. A robot often must have sensors located strategically to aid it in its tasks. These sensors are sometimes merely a switch indicating on and off signals. The usual logic networks are capable of handling this input, but often the sensors themselves are very sophisticated elements. They deliver analog signals, or sometimes, in the case of vision, very sophisticated elements of information.

As this information is received, it must be processed by one of the peripheral boards of the computer into unambiguous instruction. Some preordained task can be given to the robot based on the information of the sensor. In most cases, when a particular sensor is added to a robot, some special equipment is added to the computer allowing it to interpret signals from the sensors. Increasingly, however, as the sophistication of the base robot equipment improves, such sensory-interpretation ability is put on the computer boards themselves. This lowers the expense to the customer who wishes to add, at a later time, any particular sensor needed.

### Communication

The communication task of the robot computer is accomplished inside the computer itself and also with the outside world. Many of the signals used within the robot to control the various motions must be interpreted by the computer before they are of any use. Even the signal from one axis control to another of the same robot sometimes will be interpreted by the computer before it is delivered. Such is the case when two arms of the robot must move in harmony to produce a desired result. The signals from one arm to the other arm still will go through the computer to be interpreted.

The signals the computer receives from the outside world can take many forms. In the simplest case, they are on-off switches. Indeed, many of the buttons found on the robot console are considered by the computer to be inputs from an outside source. Even though it is the operator of the robot that is pushing the buttons, the computer does not consider the signal part of the robot. So where a great many buttons are needed to control the robot (perhaps some 100 different functional buttons), it is often necessary for the robot to employ a multiplexing system.

A multiplexing system is a way to use fewer connections and still have the same number of input lines. A multiplexer can be envisioned by considering a screen door (Figure 7.6). If we look at the number of holes in a screen door, we might find some number such as 8,000. There are not, however, 8,000 individual pieces of wire that create the screen door. There are perhaps 100 pieces of wire in the rows, and 80 pieces of wire in the columns. The number of intersecting points is similar to the number of buttons to be read on a robot.

A multiplexing system is accomplished when the robot supplies electrical power to only one strand of wire on the horizontal plane at a time. It may examine at that one instant each one of the individual intersections created by the columns of wire on the vertical plane. If a button is pushed (at the intersection of a column and row), it is only those two wires that conduct the current. The computer needs only the number of connections in the columns, plus one more for the first row. If it looks at the next set of horizontal intersections, established by the vertical wires and the second horizontal wire, it can use the same set of vertical lines as before. It needs only one signal that is different so it can interpret whether it is looking at the first set of intersections or the second set.

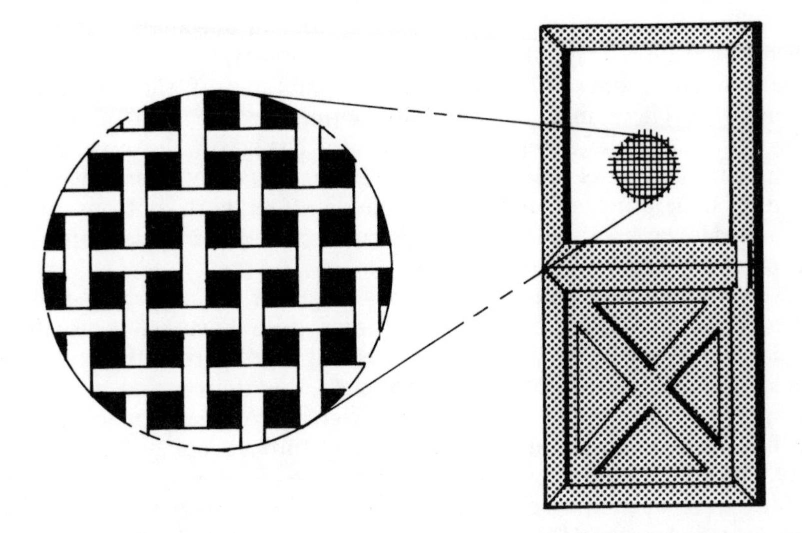

Like the intersections of wire on a screen door, the switches on a multiplexing system can be monitored with only a few lines. Although there might be 8,000 intersections, only 180 lines (100 rows + 80 columns) are needed.

**Figure 7.6.** A screen door example of a multiplexing system.

It is possible to have 100 buttons on the console of a robot interpreted by only 20 wires leading into the computer (ten rows plus ten columns). By automatically timing these selections, even greater space saving can be maintained. A microprocessor chip can be addressed with a limited number of lines (8, 16, 32, etc.). For this reason, a multiplexing system is necessary for almost all robot communications.

Another communication task of the computer is giving information to the robot operator. If there is a TV screen used by the operator, or a series of light emitting diodes (LEDs) indicating numbers or some other instructions, the computer must supply the required information on the display and keep it there long enough for the human operator to use it. In human terms this is very natural, but in microprocessor terms the few seconds it might take the operator to read the display is thousands and thousands of cycles. Usually, additional boards are used to hold up this kind of display. Even while the robot computer is doing other tasks, the individual display elements remain readable by the human operator.

The computer is also used to monitor certain automatic equipment. There may be safety or emergency stop switches located at remote points so that operators can halt the functions of the robot in an emergency. There may be automatic equipment on machinery which will either stop the equipment or cause a change in the programming when certain events happen. These automatic fail-safes are monitored by the same kind of multiplexer network as push buttons on the cabinet. However, a multiplexing system is considered inadequate for an actual safety device.

Even though the emergency stop switch for the operator may be located on the same panel as the other push buttons, it usually is hard-wired. That is, it has its own wires and is not part of the multiplexing system. It is absolutely safe to push this button and expect the computer to respond. Often, the power going to the robot motors will be interrupted when the emergency stop buttons are pushed. In this way, the communications of the computer are circumvented. If there is a failure in the computer's communication link, damage to the robot need not occur since there is equipment more powerful at stopping the robot than this communication link is at moving the robot.

Another factor of robot communication is the ability of the robot computer to use programs created elsewhere. In some robots, this is not possible. The manufacturers deliberately cause the robot to refuse any signals not created on that individual robot. In some other robots, it is necessary. The individual controllers do not possess the ability to write programs with the equipment attached to the robot. The programs must be created by some external means and then communicated to the robot cabinet. This is sometimes accomplished by plugging in a chip similar to a home video game cartridge. Often, it is accomplished by establishing some interactive communication link, sometimes using a phone line, either with a main frame computer or a larger robot equipped to write its own programs. When robots are in communication with other robots, the usual communication is not the transferring of programs. When two robots are working side by side, the communications ordinarily would be simple inputs and outputs. When *Robot A* is in position, the signal "I am in position" turns on a particular input in the other robot. When *Robot B* is ready to start its task, it will signal back to the first robot "I am ready" by simply turning on another input. Communication links between computers used for programming are generally disconnected when the programming is done.

# CHAPTER 8

## ROBOT PROGRAMMING

When robots are controlled by computers, the instructions given to a robot are in fact computer programming. Most robots have sufficient internal decoding devices to accept some human language form as their instructions. The simplest form of computer programming done on a robot is a numerical listing of the positions to be reached by the robot. By pushing a key on the robot controller, we are telling the robot, "Do this action." If we are pushing a special programming key, the instruction to the robot may be something like, "Return to this position each time we get to this point in the program." Or, it may be some subset of such an instruction, for example, the speed with which the robot is to accomplish this task or how closely the position must be maintained.

A robot program, if it were written in human language, might appear in this simple form as in Table 8.1. Here, we see a numerical listing of the individual instructions for the robot. The executive program written by the manufacturer tells the robot how to accomplish each of these steps. This program establishes which of the steps is to be done first—usually the lowest number—and, in turn, each of the other steps to be accomplished.

In some robots, a higher level of sophistication allows the instructions to take on a greater human aspect. It is possible to give language names to positions and then refer to them later in our programming. If instructions are given to the robot by typewriter keyboard, it is possible to cause the robot to move to a particular point "George" and to refer to that point as a position for pick-up. Whenever in the program we wish this point to be used, we simply use the name "George" to identify it. We can use names like "Machine 1," "Over part basket," and "Inside assembling machine" and write our programs in a very straightforward way. The actual motions of the robot and the instructions being received by the individual components are the same regardless of the programming language. By allowing a more sophisticated use

**Table 8.1. Instructions for Robot**

1. Large arm out
2. Large arm in
3. Small arm out
4. Small arm in
5. Base up
6. Base down
7. Arm extend
8. Arm retract
9. Grip
10. Release
11. High speed
12. Medium speed
13. Low speed
14. Very low speed
15. Record present position
16. Check for location of new position

Etc.

of human type language, the robot manufacturer eases the process of programming.

Some robots make use of specialized instruction unique to the physical make-up of their components. It is possible to establish on some robots a *tool center point*. This is an artificially determined point not on the robot itself, but located relative to the tool being carried by the robot. With mathematical manipulation in the robot computer, a program can be written to move this tool center point along some predetermined path without specifically programming the individual moves of the robot components.

If the tool center point is to move along a straight line, as in Figure 8.1, some complex combination of motions may be necessary in the robot arms. With the proper computer programming language as a part of the robot, the operator need only specify that the tool point is to move from one point to another. The individual arms will then move, each compensating for the actions of the others to accomplish this straight line motion. In similar fashion, very complex motions for the tool center point can be accomplished without actually creating those motions manually.

The tool center point can be programmed to move in a circular path simply by specifying where the center of the circle is in space, the radius of the circle, the starting point the tool center point will assume on

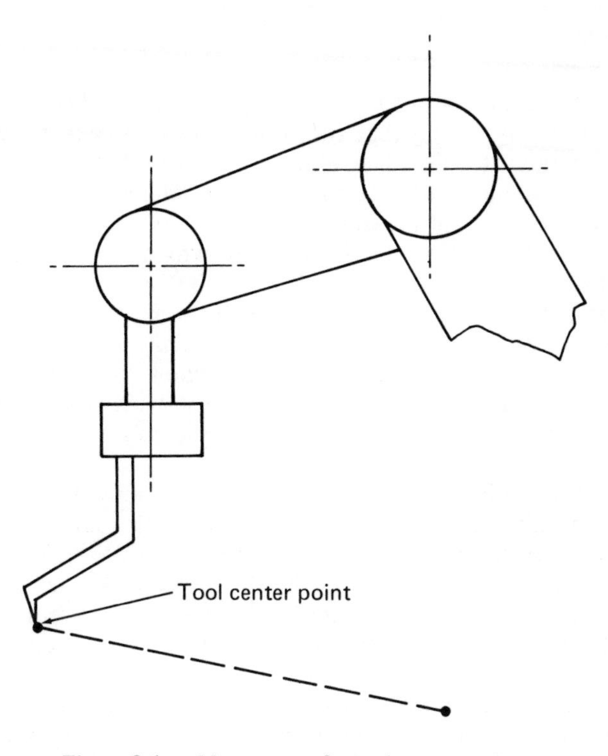


**Figure 8.1.** Movement of a tool center point.

the circle, and the number of degrees of rotation desired. The robot computer, upon receiving these signals, will compute all of the motions necessary to produce the desired motion. In even a small circular interpolation movement, there may be several thousand computations by the computer. Linear interpolation, although simpler, still presents a tremendous amount of computation by the computer. A robot's ability to perform complex tasks may be rated by examining the robot's ability to perform interpolation, either linear or circular, and the speed at which these computations can be accomplished.

In some robots, programming is accomplished by a standard computer language. If the robot receives along certain data channels a specific set of instructions, its individual components will move. An external computer can be used to generate these particular motion commands. Some standard output such as a *print* command together with a coded message will move the robot.

## ACCURACY

When we say that a robot does exactly what it has been commanded to do, we are really saying that the robot *tries* to do exactly what it has been programmed to do. The ability of a robot to do a particular task lies in its abilities to measure physical space and to maintain its position according to that measurement.

A robot's ability to measure the space in which it moves is a factor of the internal measuring equipment. If the robot is capable of measuring one thousandth of an inch, it can have an accuracy of the same amount. The robot's physical construction, however, may not allow the robot to return to the same physical position when its measuring sensors indicate it is at that position. The *tolerance* with which each of the physical components was made will affect the robot's ability to return constantly to the same position.

In modern manufacturing, many parts are made exactly the same. That is, they are made as close as possible to the same size. Because our ability to measure is very great, we can distinguish very small differences between two parts supposedly the same. We say for practical purposes that if each of the parts is no greater different from some theoretical ideal than a predetermined amount, it is an acceptable part.

This deviation from the ideal is called the tolerance. When we say there is a hole in a part with a tolerance plus or minus one thousandth of an inch, we mean, in all cases where that part is acceptable, the hole will be no larger nor smaller than one thousandth of an inch from the ideal size. The components of the manufactured robot, if made with close tolerances, will fit so as to return always to exactly the same position as their measuring devices indicate. Robot components made with large tolerances are not so able to move to exact positions.

When the robot is assembled, it is possible to measure or to compute the robot's ability to find an exact point in space and repeatedly to return to that same position. Most robot manufacturers use this repeating ability as a measurement of the robot's accuracy. If the robot can repeatedly return to some position with no greater deviation than five thousandths of an inch, it is said the robot has an accuracy of plus or minus five thousandths of an inch. In a pick-and-place robot, this position in space may be difficult to locate the first time. In a servo-controlled robot, because we have very minute control over the motions of the robot this position in space can be reached with some ease.

Depending upon the task required of the robot, high accuracy may not be necessary. If a robot is to be used for some task like spray painting, inaccuracies of one quarter of an inch might not unduly affect the quality of work. Indeed, in some designs, accuracy is deliberately sacrificed for rapid maneuvering of the robot arms and ease of programming.

In some other robots, in particular some used in assembly work, speed and ease of programming are less important than the ability to maneuver a part to an exact, predetermined position. If a cylindrical pin is to be placed in a hole and the pin is two thousandths of an inch smaller than the hole, the robot doing the insertion must be very accurate.

In some robot applications, it is necessary to control with great exactness the path of the robot. If the robot is moving a grinding wheel across some part to be sharpened, each position that it takes on the path must be held to a very small tolerance. The accuracy of a robot for this type of job depends not only on its ability to measure its position and the ability of its mechanical assemblage to yield this position, but also the ability of the computer to establish a series of these positions, in the proper sequence, and move the robot into these positions.

## FEEDBACK

When a robot is executing a program, signals must be given to the computer indicating that each partial step is complete. Only in this way is it possible for the computer to keep track of the robot's motions. The method the robot uses to report its position varies among types of robots and manufacturers. The simplest type of feedback is accomplished with a mechanical switch. Used in both pick-and-place and servo-controlled robots, a switch indicates that the end of travel of a motion-producing device has been reached.

In the pick-and-place robot, a limit switch is placed at each end of the stroke of the motion producer. When the robot executes a demand to move one of its arms, the switch is activated at the completion of the move. The robot controller will wait until the feedback signal is received before continuing. It is good practice, as the next step in the program, to examine the state of the switch on that same arm at the other end of travel. This switch should not have been activated, because the arm has moved in the other direction. If the computer should find at any time that both switches have been activated and it is apparent that a malfunction has occurred, an emergency stop sequence should be initiated.

In a servo-controlled robot, the actual position of the arms is reported by a means other than a switch; but a simple switch is sometimes used to indicate an arm is at the end of its travel. In some robots, where the motion of the arms, if carried too far, might damage the equipment, a switch can be placed along the arm indicating its position near this danger point. Upon receipt of a signal from this switch, the robot computer would stop motion in that direction. This double protection system–the normal feedback procedure plus the switch at the end of travel–allows the robot to be self-correcting, even if a program for manual control by the operator commands the robot to a position where it would damage itself. The computer would not allow the completion of that command.

Another device used primarily in servo-controlled robots to indicate position of the robot arms is called a *resolver*. A resolver is basically a series of electrical windings. Two or more windings are placed in a housing. An additional winding is placed on a shaft that will rotate within the housing. As some type of pulsed electrical signal is provided to the housing windings, a characteristic signal will be developed on the shaft winding that will change as the shaft rotates. If the housing windings do not move and the signal given to those windings does not change, the exact position of the shaft as it rotates can be measured. The device is said to resolve a portion of rotation of the shaft when it changes the shaft signal sufficiently to be measured. It is common for these devices to measure in the range of one degree of shaft rotation. This small amount of motion allows for great accuracy in the positioning of the robot.

A resolver, however, cannot indicate the position of a robot arm if the motion of the arm exceeds 360 degrees of rotation. Each position on the resolver indicating a certain number of degrees of rotation is exactly the same, no matter how many times the shaft has been completely rotated. For this reason, when a resolver is used, it must be coupled with some other measuring mechanism. Usually the computer will count the number of times a resolver goes through one complete cycle of rotation. A combination of these two pieces of information–the resolver position and the number of complete resolver rotations–indicates a course and an exact relative position for the robot.

An *encoder* is a device for measuring rotation of a shaft. An encoder can measure extremely fine pieces of motion. However, each increment of motion, once counted, gives exactly the same signal as each other piece of motion. There is no indication from an encoder as to the actual position of the shaft.

Encoders are commonly assembled as an interference arrangement of small light passages. A series of lines can be drawn on a mirror which remains stationary with the body of the encoder. A similar series of lines, as shown in Figure 8.2, can be painted on a transparent disk that rotates with the shaft. A simple light source shining against the mirror can be detected by a photoelectric cell. As the transparent disk rotates, it will alternately occlude the light reflected from the mirror, and, when the open areas between the painted lines on the mirror and the disk are aligned, transmit all of the available reflected light. The pattern of lines painted on the surfaces can be extremely fine.

The pulses of light or darkness can be counted by the robot computer

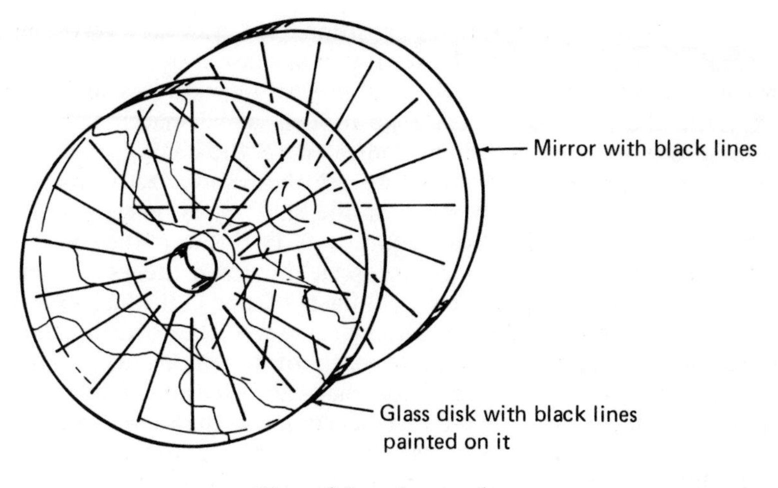


**Figure 8.2.** An encoder.

and used to indicate motion of the arms. Because the signal from the photocell is electronic, it can be very rapid. There is no contact between the stationary and rotating elements of the encoder, so there is virtually no limit to the speed at which an encoder will function. It is only the mechanical ability of the last rotating disk to withstand centrifugal effect that limits the speed at which the encoder is turned.

Encoders can indicate extremely fine motion of their shafts. An indicating range on the order of one thousandth of a revolution is possible. If a robot using an encoder as its principal measuring device can always count the encoder pulses for each move made by the robot, theoretically the encoder will always remain correctly positioned between the program and the physical robot.

Encoders are more accurate than resolvers. However, because they indicate only incremental motion, they are less able to aid the computer in determining some particular position of the robot, for example, when the robot control has been turned off and we wish to start the robot.

Some home position is usually sought when starting the robot. The robot arms are manually positioned to some starting point. It is common to use a series of switches or artificial clamping mechanisms to achieve this position. When an encoder is used to indicate this starting point and all motions away from this point, the switches and clamps establishing the starting point must be very exact in their own positions. If there is any deviation over time in the position of these switches, the encoder is not capable of indicating this deviation. The robot programs written relative to this original starting position will now at every step be inaccurate by the amount of the starting-point deviation.

When a resolver is used—with its ability to indicate in which portion of a circle the shaft is currently positioned—small inaccuracies of the starting-position switches will be indicated to the computer. The computer may be programmed always to use the resolver zero position—or some other characteristic portion of the shaft position—as its starting point rather than the actual switch position. In this way, the robot home position is established on a course scale by the switches and clamps and on a fine scale by its resolver. Each of the individual axes must have its own measuring device. The inaccuracy of the overall robot is a summation of the inaccuracies of each of the individual arms. The small deviation in the home position that can be corrected by a resolver system can make for greatly increased accuracy in the robot itself.

When possible, some robots use an *absolute measurement system*. In an absolute system of measurement, the measuring device indicates a unique signal for each incremental position along its entire measuring range. Many machine tools with automatic measuring capabilities use a glass slide as a means of measurement. The glass slide is really a tube with very fine windings allowing a moving part outside the tube to indicate its position constantly.

In most robots, the use of an absolute encoding system is not possible. The position-producing device is located in such a way that the portion of the robot making the greatest movement is not available for the placement of the measuring device. The small motion made at the robot point convenient for measuring motion allows for great inaccuracy if an absolute positioning is used. Robots that have motion produced along some Cartesian-coordinate positions, however, are ideal candidates for an absolute positioning system. A strong advantage for an absolute positioning system is that inaccuracies within the robot itself will not cumulatively change the position of the robot as it tries to reach some artificially established point.

If the robot command is to travel, in one axis, ten inches, an absolute measuring device allows, within its own ability to measure, for an exact placement of the robot arm ten inches away. In some type of shaft rotation measurement system (despite great effort by the robot builder and designer), a computation must be made by the computer as to the number of shaft rotations equal to ten inches because the exact distance a robot arm will move in one rotation may be slightly inaccurate. This inaccuracy, added to for each rotation, would be sufficient for a great inaccuracy after many rotations.

In addition to indicating position of the robot arms, feedback signals are also used to control other functions of the robot. By measuring the weight of signals from an encoder or the rate at which the signals of a resolver change, the speed at which a robot arm is moving can be computed. In most robots, as a motion is begun a computer will allow the arm to accelerate gradually. Constant updating of the speed of the arm allows the computer ever-increasingly to set the power to a high level. When the robot arm nears completion of its motion, the computer allows the arm to decelerate. A continuous check of the arm speed allows for an increased deceleration, if necessary, to stop the arm.

If the robot is carrying a heavy load, a greater change in power may be necessary. By examining the speed of the arm as it is decelerating, the computer may make new calculations as to the needed deceleration force and act accordingly. In some robots, additional devices are added to indicate speed, particularly where a resolver system is used. The amount of computer time spent in calculation of speed can be reduced if some additional equipment is added giving a simple reference output of speed. This can be accomplished with a simple *tachometer*. The tachometer can be used both as a speed-measuring device and, when used with the computer's own timing ability, as a double check to the robot's position-measuring device.

Another function of the robot feedback loop is to indicate when the arms are moving spontaneously. If the robot program calls for the arms to be at a particular point and to remain there, it is the feedback system that signals a robot control if the arms begin to move. Indeed, while an arm seems stationary to a human viewer, it may be moving along one of its axes hundreds of times per second. Occasionally, if the feedback system and the motion-producing system are out of balance with each other, there is no position that the robot will reach that will satisfy the feedback system as being a correct position.

When the motion-producing devices stop their motion, this may appear to be one increment of motion too far to the measurement system (Figure 8.3). As the computer commands a retract of the one-measurement position to the motion-producing device, the actual distance moved is enough to cause the measurement system to indicate the exact position of the arm as one increment in the negative direction.

This process will reverse itself and then repeat as often as the computer analyzes the current position of the robot arm. At these times, the arm is said to be *dithering*. In some cases, dithering is a desired phenomenon and is produced on purpose. Unless it is designed as part of the system, however, dithering can be very wearing and even destructive to robot components.

Another element of feedback control is the robot's monitoring of its own power consumption. In a hydraulic robot, when the robot's arms encounter some obstruction the robot will try to push through the obstruction. By monitoring the amount of pressure a given motion usually takes, we can establish a sensing mechanism (in this case a

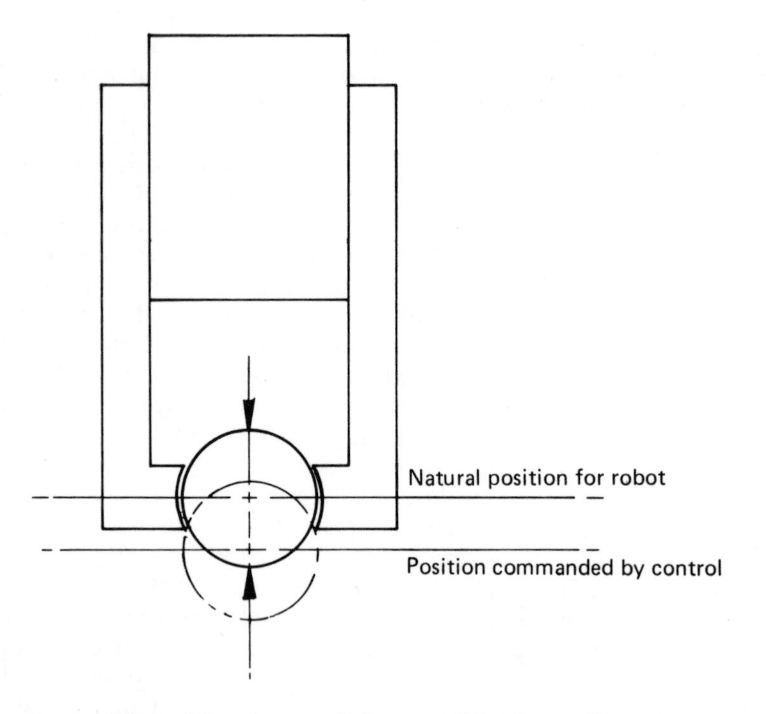


**Figure 8.3.** An out-of-phase condition for resolvers.

pressure switch) that would indicate we are consuming too much power for this given action. It is possible now for the computer to institute an emergency stop signal before the robot damages itself or surrounding equipment.

In an electric robot, the current supply to each of the electric motors can be monitored. If too much electricity is being consumed for a particular operation, a measuring device can indicate this condition and the computer can stop the motors. In some robots, a compensation is made at the trigger point of this overload sensing so that heavy loads may be handled without triggering the setting. Yet, when light loads are being handled, even a small obstruction to the robot can signal the computer.

# CHAPTER 9

# ROBOT SENSORS

## ROBOTS THAT SEE

It has long been the dream of robot designers and science fiction writers that robots gain the ability to see. Some robots in use today make use of real vision systems. Vision can be as simple as a single point of light or as complex as a color TV picture. However, all vision systems, regardless how complex, use single points of light as a source for their sensing ability. Although points may be combined and sophisticated analysis applied to the combinations, this single piece of information forms the basis for all robot vision activity.

### Beam Detectors

The simplest form of robot vision system is a photocell which measures the presence or absence of light. When a robot is configured so a continuous source of light and a detector for this light are aligned on the arm of the robot, the beam detector can detect the presence of a part. The robot maneuvers the beam detector into a position and then checks to see if the beam has been interrupted. In this way, successive moves can be made until a part is found.

If the beam and its photo-detection target are very small and accurate, the robot is capable of finding small pieces. If, however, the light source or the detector has large areas, a small part or other intrusion of the beam could go unnoticed. Beam detectors are useful in the manufacture of continuous strips of material. If a beam detector is aligned so that a long, continuous piece of sheet metal will break the beam, we always can be assured we have not reached the end of the sheet as long as there is no input signal to the contrary.

### Light-Level Detectors

Beyond the simple presence or absence of light, a detector can sense the level of brightness. A light-level detector can, for example, inspect glass sheets or the reflective properties of material being polished. If we design a robot program to polish a metal surface until its reflectivity reaches a certain point, a light-level detector can signal the robot when its work is done. The detector in this case does the actual analysis of its input signal without the aid of the robot computer. When the light level reaches the proper point, the robot receives the command signal from the detector, "You are finished."

Light level detectors are somewhat inadvisable for the industrial environment because the nature of most jobs allows for contamination of the sensing device. To continue with the polishing example, polishing powder or grinding dust may occlude the detector's surface, thereby reducing the effective signal of the reflective lights. This may give a false signal to the robot. With the surface at a perfectly polished level, the robot may not receive its signal to stop and remain polishing forever.

### TV Cameras

It was thought early on in the use of robotics and other automated equipment that a vision system could be established merely by connecting a TV camera to the robot's computer. However, the process is not nearly so simple. When we view a TV screen, the image we see is very simple for our human brain to understand. The image on the TV screen, however, is usually of no value to a robot, unless it is analyzed by special *algorithms*. First, let us examine how the human eye sees and compare this to a TV camera, and then we will consider how a robot can use the visual information collected by a TV camera to perform a sensing function.

The human eye is composed of three major elements: a lens, which is the clear, curved-shape portion in the center of the eye; an iris for variable opening; and the actual image light-detecting nerve cells within the iris. When light strikes some object in front of the eye, as seen in Figure 9.1, the light is reflected from the object and passes to the eye's lens. Because of the lens' shape, the light from the entire eye surface is focused on the small area within the aperture of the eye.

This focusing process allows the eye to see the image clearly, even though the light reflected from the true surface is continuously spreading out as it leaves the object. If the light is too bright, so that it would damage the eye, the aperture opening will close slightly to reduce the overall input of light. After the focusing process and after some portion

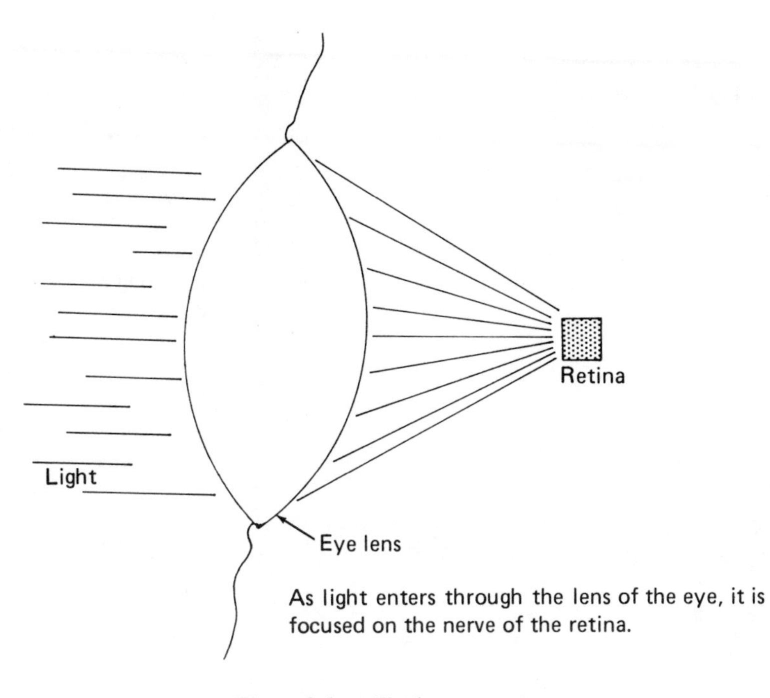


**Figure 9.1.** The human retina.

of the focused image is deleted by the aperture, the light falls on the iris and can be detected.

The iris is composed of a bundle of individual nerve cells. Each of the cells has the ability to sense both the level of light, and, in combination with the other cells, to sense the color of light. It is really a collection of individual segments of light which we see. The nerve cells are extremely small, and therefore the resolution of the human eye is very fine. Our ability to see a small object is based on the relative size of these nerve cells. Information recorded by the cells is transmitted along the nerve lines directly into the brain where it is analyzed.

A baby can recognize its own mother's face and show a preference for it. Although we consider this a natural phenomenon, it is by comparison to the robot vision system extremely complex.

A TV camera has many similarities to the human eye. There is a lens used to collect the scattered light of an object, as seen in Figure 9.2, and a shutter device to limit artificially the amount of light allowed to

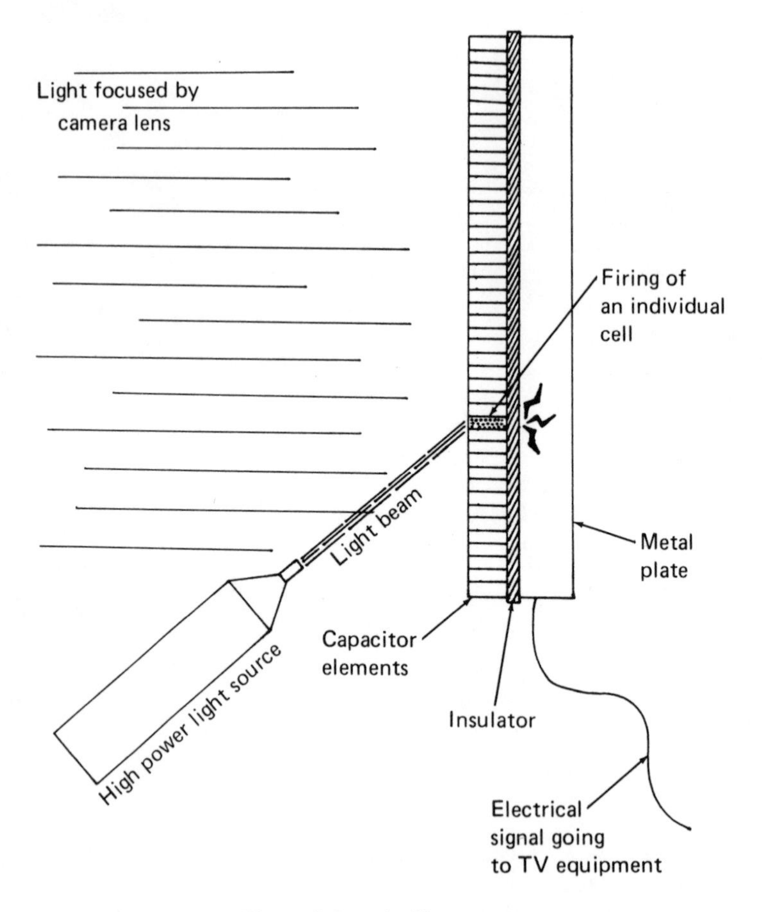


**Figure 9.2.** A video camera.

enter the camera. Rather than allowing the light to be detected directly by sensors, a TV camera collects the light and stores it as electrical information. A plate inside the camera has thousands of tiny photo-sensitive elements. All of these elements are placed on a nonconductive sheet, and the sheet is placed on some metal surface.

As light is allowed to enter the camera, very minute sections of the image upon which the camera is focused are indicated by the photo-sensitive cells. Because of the nonconductive nature of the sensor, a small capacitor element is produced. Various levels of light on each of the small cells create varying charges on this capacitor. These charges are small enough that, by themselves, they do not dissipate.

Another element of the camera is the means by which the information stored in the cells is removed. A very powerful yet small beam of light is directed at the cells, one at a time. As each cell receives this intense beam of energy, the capacitors' and insulators' ability to maintain the charge is lost. Electrons migrate to the capacitor from the beam and from the capacitor through the insulating wall to the metal conducting plate. As this energy is conducted away from the plate to some desired use of the signal, the beam will focus on the next cell. We have a series of impulses which can be analyzed in exact order and later assembled to reform the image.

When a television signal is broadcast to a home set, it is this series of individual images which the set receives. Because certain standards have been established for the number of capacitor cells in the camera and the rate they are sent, all television sets can receive an image point by point. At the television set, individual dots of fluorescent material inside the picture tube are signalled one by one for the desired level of light they should produce. Proper arrangement of the lines of dots in the television set allows the original image to be duplicated at the remote point.

When a TV camera system is used in a robot, the image seen is the same as the raw signal from the TV camera. If we wish to perform some sophisticated vision process (Figure 9.3), we would now analyze

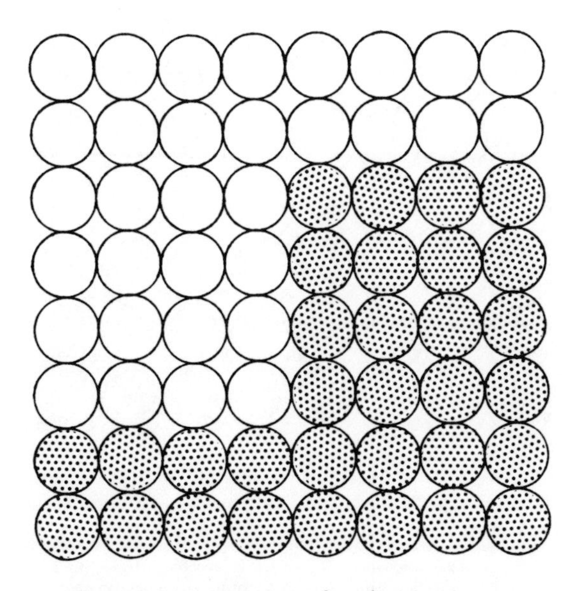

**Figure 9.3.** A portion of a television image.

a long string of individual pieces of information. In a normal TV camera image, there are many thousands of individual cells to analyze. If we consider the completely random elements viewed and allow for deviation in perceived size due to angle and distance variations, it becomes obvious it is a very difficult task to find the single part enmeshed in the tangle of lines available to the camera eye. It is possible, however, to use the camera in different ways and yet make use of the information with simple analysis.

It is not always necessary, when using a TV camera, actually to detect the presence of an individual sub-image located randomly in the field of vision. Often, only some moving element within the field need be detected. If we are grinding (Figure 9.4), the operation may be complete when a certain amount of material has been removed. If we store in memory the total number of points of a light level indicating the solid shape of the part, we can then continuously check this number against the new information of the part size as it is being ground. When the actual number of points has changed by a predetermined amount, the vision system can recognize that the grinding operation is

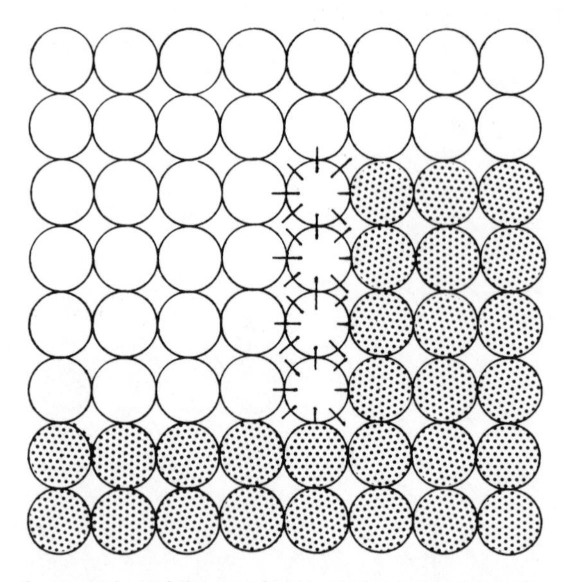

As a portion of the part is removed by a machine, the camera image changes and will trigger a response from the robot.

**Figure 9.4.** The changing of an image on a television signal.

complete. It is not the actual shape of the object that is being measured, only its change from a prerecorded piece of information.

Similar to the analysis of counting a number of cells is the ability of a camera to find and orient a single part. If we have a conveyer and only one part may be expected at a time, it is possible with relatively simple analysis to determine an edge of the part. The edge may be defined in computer memory as a certain combination of points, all indicating a solid component. The computer may analyze, in an orderly fashion, small segments of the image and determine if they fit the requirements for a line. Once the line is established, the orientation of the camera relative to the robot gripper will indicate the motion necessary to grip the part properly. Thus, randomly placed parts can be oriented by the robot for placement in some other machine.

If a simple analysis of a TV camera image is not sufficient for the job at hand, there are systems that can truly see with their analysis algorithms–such as an algorithm that allows a single complex image to be resolved (from a field of vision containing many similar images and much extraneous information) and that can be used to direct the path of a robot to pick a single part out of a parts bin.

The analysis for such vision signals is usually not a function of a robot itself. Rather, it is a special piece of equipment that is self-contained and, like its simpler cousins, uses only a series of commands to the robot. Once the target object has been located in three-dimensional space, a series of commands is given to the robot. These are to orient the arms and gripper to prepare for pick up. By adding and subtracting images, comparing minute portions of the overall field, and artificially adding segments of the field which appear by angle or color contrast to be missing from the proper image, as it is described in memory, the vision computer system locates the object.

If the camera itself is sometimes mounted on the robot with the gripper in its field of vision, changes in the robot and in the part can be computed interactively. If the camera is mounted in some stable position and the robot is not in its field of vision, a continuous orientation in memory of the relative positions of the camera and the robot must be made.

At times, it is the job of the robot not to manipulate some part as seen by the camera, but rather only to carry the camera to different positions. This is the case where robots are used for inspection of goods. Here it is often unnecessary for the robot to be in communication with the camera system. The camera can go about its sensing duties while being carried by the robot, which is completely ignorant of the task being undertaken. Some third piece of equipment may be storing the individual dimensions of the part as measured by the camera, but this information may not be necessary to operate the robot.

Other equipment is sometimes used in conjunction with or in place of a regular TV camera. Fiber optic connections, as seen in Figure 9.5, can be used to route the visual signals along complex paths, sometimes within the body of the robot itself, before the signals are analyzed. At times, special light sources are used that are beyond the optical senses of the human eye. Infrared light sources are sometimes used, as is laser light in the ultra violet spectrum. The sensing device may be some equipment like an X-ray machine analyzing a component for defects. It is only the end signal the robot is concerned with, and any technology that can be configured to provide the signal is a candidate for use in a robot vision system.

The majority of robots in use today and planned for the foreseeable future will not use vision systems, even though they are available. Two parameters act as constraints for the use of robot vision.

The first factor is the expense of the equipment itself. It is sometimes more expensive to purchase the vision system for a robot than the robot itself. If the image to be processed is sufficiently complex, a vision system is extremely expensive and sometimes is only useful for its

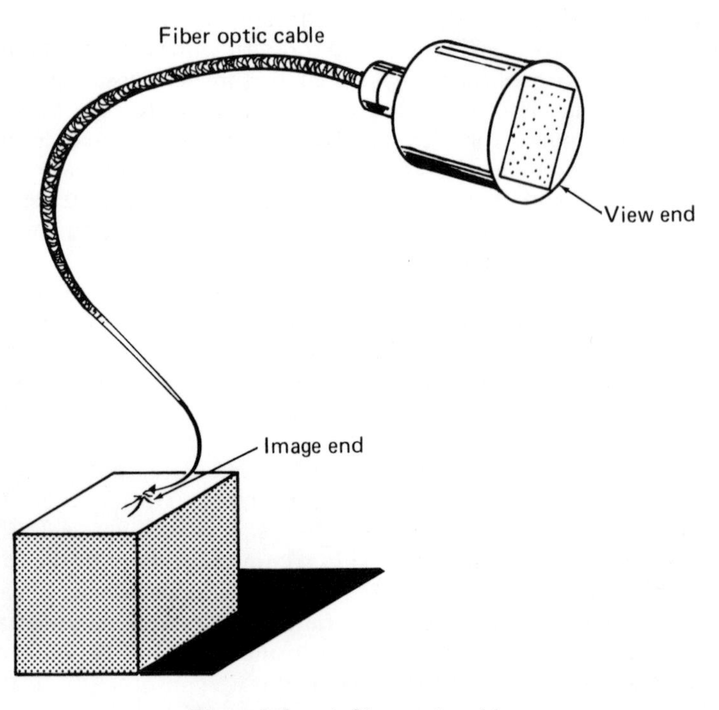


**Figure 9.5.** A fiber optic cable.

designed purpose. It is usually cheaper to orient the robot with some means other than a vision system. If some mechanism can be found to orient automatically the parts given to a robot, the robot need not have a vision system to do this operation. As in the example involving a complicated network of parts, the overall cost of a vision system to orient these parts is much greater than a machine to orient them in a position suitable to a blind robot.

Likewise, the second constraint on robot vision is usually sufficient to disallow the use of vision as a practical tool. Robot vision systems slow the work of robots. If a program can be written simply to move the robot to a predetermined point where it can carry out its functions, no time is lost. Time is lost in the orientation necessary for a robot to perform the same function with a vision system.

As more robots are used in industry, it is becoming less necessary for a robot to orient parts continuously before they are processed. Inherent in automation, a part is oriented just before it is released by the robot. In a proper scheme of manufacture, the orientation as a part leaves one robot can be used as the orientation for the part arriving at the next robot. Part-handling magazines and conveyers can maintain the orientation from one work cell to another.

The greatest use of vision at this time is in continuous path operations where a part itself may deviate in its relationship with other components. If two pieces are to be welded by a robot, as shown by Figure 9.6, the relative position of the parts may change by factors beyond the control of industry. A properly designed vision system that can detect these

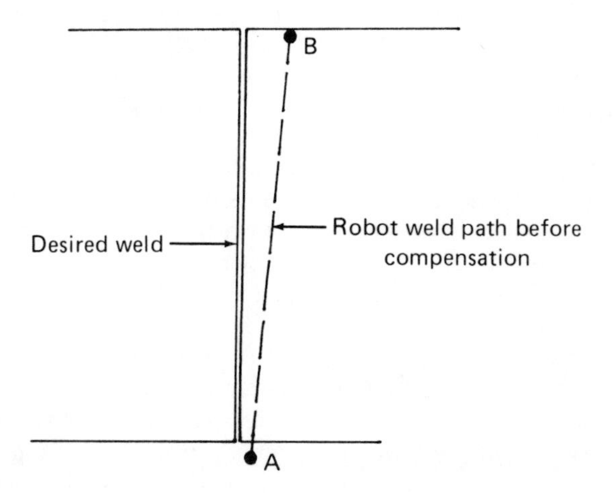


**Figure 9.6.** Example of weld compensation by robot vision.

changes and, while the robot is continuously moving, alter the robot program to compensate for the changes, is an extremely useful tool. While the ability of robot vision sensors is far from that of even the simplest human vision, vision has reached such a state of art as to make it a viable sensing medium for the modern robot.

## ROBOTS WITH A SENSE OF TOUCH

Very often in industry, robots use the ability to touch an object and gather information from this touch.

### Mechanical Switch

The simplest form of touch sensor is a mechanical switch. As the robot arm moves through a preprogrammed set of points, the robot controller continuously monitors the switch to check for an input signal. When an object is found, the switch signals the controller by conducting electricity. A program may be written so the robot, upon the trigger of the switch, can pick up the object or do other work from that relative starting point. If an object is to be located in space–not along a particular line–the path the robot takes in seeking the part can be a series of more complex lines, such as the spiral.

A drawback of this sensing type, often precluding its use, is a tendency for the robot arm to be mispositioned after the switch's signal is received. Two factors involved are: servo lag and multimicroprocessor *scan rate*. At the exact moment the switch is triggered, the arm is moving through space. Although it is at the correct position and a signal from the controller may reach it, and although a stop signal may reach the arm very quickly, the laws of physics require some deceleration time to bring the arm to rest. Although the computer program may be sophisticated enough to allow the arm to overshoot its position and then return to the correct point at some lower speed, the arm still, for a time, has deviated past the stop position indicated by the switch. At times, this servo lag is sufficient to crash the arm of the robot.

The factor of scan rate is smaller in scale, but of greater significance in positioning accuracy when a switch is used. The way most microprocessor components are configured, the input channels are analyzed at only certain clock cycles. The signal from the switch being used as a sense of touch is most likely not entered into the computer as an interrupt signal. Rather, it is a signal like any other being read as part of a multiplexing system. At any particular moment, the microprocessor may or may not be examining the state of the particular switch in use.

Any time the microprocessor spends scanning other information is time in which the arm will move away from the correct position.

As mentioned, it is possible with servo lag for the computer to retrace its steps and compensate for error. However, when the microprocessor, by the nature of its configuration, does not witness that the switch is closed until some later moment, the information is lost as to the true position of the touch and no compensation can be made. If the robot microprocessor is powerful, in that its functions are carried out with great speed, the amount of error induced by this scan rate should be very small.

To make a realistically accurate approach to a part, allowing for these two types of error, the robot arm must move slowly. As was the case with vision systems, even though a switch controlling the sense of touch is technologically possible, the economics of its use based on slow speed oftentimes preclude its use.

## Force Feedback Signals

Another way in which a robot achieves a sense of touch is by a *force feedback* mechanism at the end of the robot arm. Here, it is indicated not merely that the arm is touching or not touching an object, but rather the *pressure* on the object. By an arrangement of elastic elements, it is possible to have a small feedback signal when the arm has just barely encountered the object and a large feedback signal when the arm is pushing more heavily on it.

This arrangement allows the microprocessor to monitor the touch device while the arm is moving quite rapidly. As the initial signal is received, the arm can be decelerated quickly, and the sensor element can report new information as to the arm's position. The object can be sought and found without overshooting the position or having uncorrectable error based on the scan rate of the microprocessor.

A type of force feedback mechanism can also be used interactively to change the robot program with changes in the parameter of the part itself. If a particular operation such as grinding (Figure 9.7) requires a force of five pounds against the part, the robot can continuously monitor the sensor to achieve this force, even though its relative position to the part may be constantly changing. As the grinding wheel removes pieces of metal from the part, the position of the grinding wheel in the robot arm continuously changes. The amount of force being exerted against the part by the grinding wheel remains at the predetermined five pounds without stalling the motor.

A particular type of force feedback unit is one using a *strain gauge* as its central element. A strain gauge is usually configured of some solid

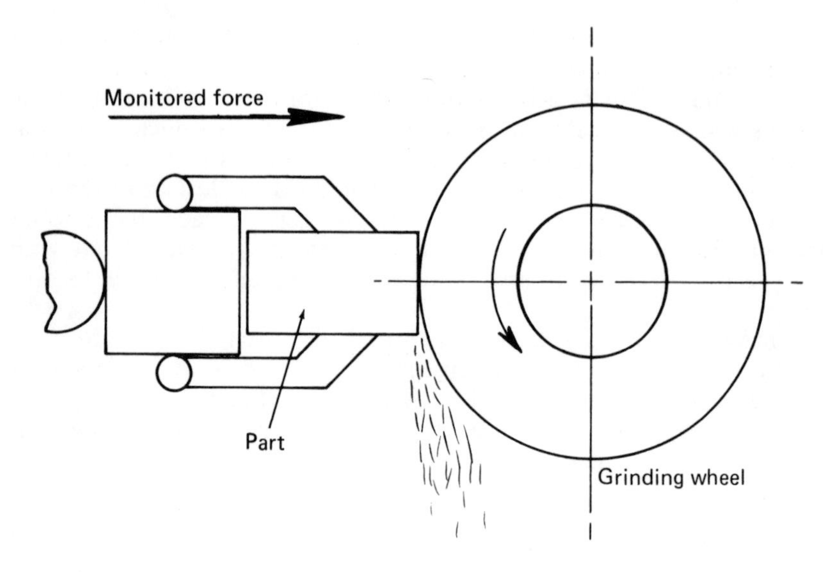


**Figure 9.7.** Force monitoring in a robot.

material which gives off a characteristic change as it is deformed. There are used in industry optical strain gauges, sonic strain gauges, and other types. The type most often used by the robot industry is a strain gauge based on a crystal material which gives a characteristic electrical change as a physical force is applied to it. A crystal strain gauge is an ideal component for a robot sensing device, in that it is very small, can be physically located within some other component, and gives a signal quite readily to a microprocessor component.

Some robot components themselves can be used as touch feedback units. When a robot arm is moved, the actuator producing the motion uses some energy. As the energy is consumed, it may be measured and the information compared to some standard. If a hydraulic robot arm is pushing against a bearing with 800 pounds of thrust, a pressure-measuring device on the input of the hydraulic actuator can signal the microprocessor of the pressure being used. If the arm has not pushed far enough to seat the bearing, the pressure will remain relatively low. When the bearing is partially seated and creates resistance that is still less than 800 pounds, the sensor will indicate the process must continue. When the bearing is fully seated and at least 800 pounds have been exerted for some time, as the robot actuator continues its motion the pressure will rise. The sensing device will now signal the computer that the operation is complete. Although the arm of the robot may possess enough strength to damage the bearing if full force is used, the sensing mechanism can be used to protect both the robot and the part being assembled.

A similar arrangement can be made using electrical and pneumatic components. In an electric robot, the current used by the electric motors can be monitored. If too much current is being consumed, the level can be limited by the controller. In a pneumatic robot, if the pressure on a particular cylinder is too high, it can be changed if proper devices are installed and air pressure is adjusted to the proper level.

A limitation of all sensors incorporated with a robot itself is their ability to compute the desired phenomenon, which would be measured based on the information available as part of the robot. The exact force produced by a hydraulic cylinder, although theoretically based only on the input pressure of the oil in its actuator, may in fact be quite different than its measured level. For this reason, robot element sensors are usually used only as a protection device for the equipment itself rather than as a sensing device for the process.

### Proximity Sensors

A robot does not actually have to touch an object to sense its presence. A robot can detect the proximity of a part by using a special family of sensors. Generically called *proximity sensors*, these devices measure electrical and magnetic fields, atmospheric air pressure, sound waves, and the presence of high energy particles.

One simple type of proximity switch is the *Hall effect transducer*. The phenomenon referred to as the Hall effect was discovered over 100 years ago. It is basically the intervention of a magnetic field on a metal plate and the effect this intervention has on an electric current passing through the plate. Hall effect sensors are used to detect only the presence of a magnetic field. It is possible in industry purposefully to apply a magnetic field we wish sensed. As seen in Figure 9.8, a pin located on a parts-carrying conveyer can be magnetized. If the part to be worked on by the robot is in some way magnetic, it is simple for the robot to locate the the part with a Hall effect transducer.

Other types of proximity detectors are used to detect the presence of metal objects. Some types use an interrupted magnetic field as their basis for sensing. As the sensor approaches a ferrous or other magnetic metal object, the interrupted magnetic field is completed by the presence of the metal. An internal device is triggered by the enhanced magnetic field and can signal the robot of the part's presence.

In some proximity detectors, a magnetic field is interrupted by the presence of a nonmetallic, yet solid, object. The same triggering mechanism is used as before, but in reverse.

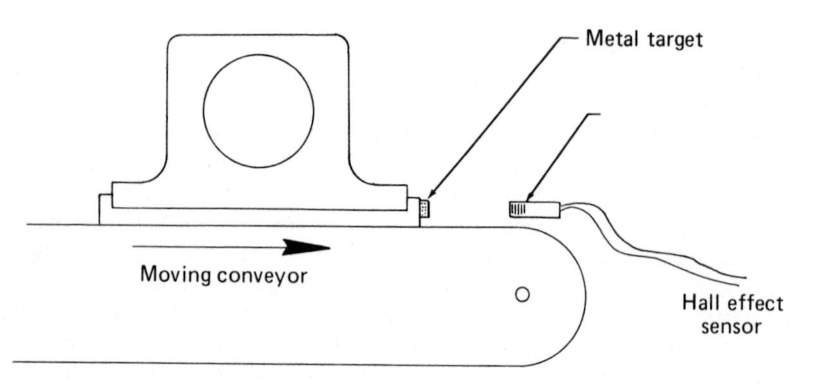


**Figure 9.8.** A Hall effect sensor.

These sensors are limited by their inability to sense all types of materials and, therefore, have a tendency to mis-trigger if presented with an unusual part. Oil or paint on a metal object can blind the detector to the presence of the metal. A thin coating of magnetic-type metal dust can falsely signal a nonmetallic sensor. In many industrial plants, the presence of contamination in the atmosphere and as part of the working cycle oftentimes precludes the use of these sensors in robotics.

A blast of air given from a properly designed orifice can be used as a proximity-sensing device. Air pressure in the chamber immediately adjacent to the exiting orifice can be monitored by a sensing device. As the air blast is impeded by the presence of some solid object, the air pressure in the chamber will rise. The closer the object or the greater the surface area of the object, the greater the chamber pressure. When a very small space is available for the robot to indicate the presence of an object, or where the object must not be exposed to electrical signals, an air-blast proximity detector can be of great use. Because there is a continuous stream of air through the orifice, it is difficult for contamination to find its way into the sensing chamber. Practically speaking, however, when such a device is turned off during periods of no production, contamination can find its way through the orifice and into the sensing chamber. If the orifice—usually quite small—becomes blocked or occluded, the sensing device may erroneously indicate the presence of an object.

Similar to the air blast, reflected sound waves can be used as a measuring medium. This *sonar sensor* can be used to measure points inaccessible to the robot. The characteristic response of certain materials to sound waves and the reflectivity of the materials' surfaces allow a sonic-measuring device to gauge the thickness of metal when both sides cannot be reached simultaneously by the measuring device. Sonic feedback, when working at extremely high frequency, is usually safe from the noise in an industrial environment.

When many other proximity-detection devices fail, it is sometimes advisable to use a *nuclear beam* as a sensing element. Much like a light beam, a nuclear beam can be detected by a sensor. When the sensor indicates the beam is not present, the intervention of a physical object is indicated. Nuclear beams, because they involve very high-energy particles escaping from the decaying nuclear matter, can be used in environments were light, sound, and pressurized air are not able to penetrate. In an environment so clogged with soot that human vision is virtually useless, a nuclear beam is pervasive enough to penetrate the dust and sense the presence of an object. The amount of nuclear material needed for such a beam need not be large. Levels of radioactivity small enough as not to be considered a human health hazard—even after years of exposure—still produce characteristic signals of high resolution in properly designed detectors.

A special type of touch or proximity sensor is the *guide wand*. When a robot is to follow a certain physical feature on the part on which it is working, a guide wand may be used. The wand is placed at the end of the robot arm close to, say, the welding tip or glue gun being used by the robot. If a certain groove is the position desired for the robot action, the wand can be programmed so that any deviation out of the groove will push against the wand. The wand will indicate to the robot controller an out-of-position signal, and the robot program will automatically return the robot to the proper position.

Sometimes the wand is part of a completely separate mechanism mounted on the end of the robot. As the robot carries out its program, the wand and its associated equipment will independently compensate for small deviations from the desired task. Wands, because they are used in close proximity to the work in progress, must sometimes be considered expendable items. If they are placed too close to an arc welding tip, for example, flash and debris from the arc may, over time, destroy the wand. With careful design, however, the tip can be made of an inexpensive material and removable.

## ROBOTS THAT HEAR

Recent improvements in microprocessor technology have allowed the development of real time artificial speech recognition. A computer can listen to the speech of humans, correctly identify the words, and respond to them.

## User-Defined Words

In some *speech recognition units*, the individual user identifies the words and their meanings. By speaking into a microphone while simultaneously giving an electronic signal to the speech recognition unit, the user allows the computer to memorize certain words. The words are each pronounced several times so that minor variations in a particular voice will not cause the computer to fail to recognize the words.

Some units that do not have tape or nonsolid-state components have only the ability to recognize a few user-defined words. Generally, however, the actual storage space for the information necessary to recognize these words is stored on *electronically programmable read only memory* (EPROM) chips. Chips created by one user can be removed and replaced by the chips of a different user.

In speech recognition units taught by user language, the robot usually will only respond to the original speaker. The word *cat* spoken by one person is treated as a completely different symbol than the word *cat* spoken by another person. There are advantages to this scheme when applied to the control of machinery. Just as a properly trained dog will respond only to commands from its master, a robot hearing words from some person not in control of the robot should not respond. By the use of user-defined speech recognition, this is possible. The intended robot programmer can carry with him a set of previously programmed EPROM chips, and, when he desires to program the robot, can install these chips. The robot will respond to this person only. Workers or visitors walking by and casually uttering the word *cat* would not catastrophically cause some failure or action in the robot.

## Responds to Language

Voice actuation units that respond to human language—in much the same way as a human—are independent from individual speakers. These units are usually programmed by the manufacturer to respond to individual phonemes, regardless of variances in the individual speaker. Such a system usually must be exclusive enough so that some words, if spoken slightly irregularly, will not be accepted. However, a 98 percent rate of acceptance can be achieved.

The advantage of these units is that anyone can program the robot upon learning the necessary key words. A password can be devised so an untrained person could not program a robot inadvertently. Because the voice actuation system responds to any voice saying the proper words, the programming can be arranged to accept command words only after the password has been received. A nonsense combination of words can be used at the beginning because it almost never would be spoken randomly by someone not trained in its meaning.

In considering the two voice-actuation systems outlined, it should not be assumed it is common practice to use these systems. The industrial environment is full of harsh and unwanted noises. Although voice-recognition units can listen so selectively that unwanted noise will pass through undetected, it is still necessary to have some degree of quiet to use voice actuation. Other workers in the robot's area might utter exactly the sounds that would cause a response.

### Double Entry System

To avoid that contingency, a *double entry system* can be devised. In such a system, the robot responds to its programmer each time a voice-actuated command is given but an additional response from the programmer is needed before the program action can be carried out.

## ROBOTS THAT SPEAK

Robots speak in a different way than humans. The human mouth, tongue and vocal chords form a mechanical instrument to produce speech. When the human brain desires to say a particular word, the actual commands given to the muscles are quite different than the commands a robot computer uses to produce sounds on a speaker. As a human begins to produce a particular word, changes are made in the position of the jaw, the spread of the teeth, the placement of the tongue within the mouth, and the way the tongue allows air to pass over it. The muscles controlling the vocal chords tighten or loosen them to cause changes in pitch. The lips, at the proper time, close or open to give emphasis and delineation to the words.

The "talking robot" has no such mechanism available to it. Rather, it has, in a look-up table, a series of electrical modulations that simulate the action of the human body parts. By correctly manipulating these electrical modulations, the robot computer forms words. Just as in human speech, sounds are joined into words. And, words are strung together into sentences with articulation and intent considered. Pronunciation of an individual word changes by the word's meaning and its placement in a sentence. Although robots can be programmed regardless of the accent their voice response might have, it is a great benefit to have a robot speak in a natural human fashion.

Speech actuation units and artificial speech production units are readily available today. Aside from the interference in the industrial environment, there is one great reason why robots are not often voice programmed. When a robot is programmed by the push of buttons and when the program is displayed on some type of operator-viewing screen, there is reasonable assurance the desired program is being used by the robot. This is not so much the case with voice actuation. We can see in our everyday lives that most people believe the written or mechanically stored message is more powerful and accurate than the spoken message. An order placed verbally with a store often must be confirmed in writing. No bank will allow only a verbal command for withdrawing money. Some physical system must be used in the communication. This same bias against voice actuation accounts for the disinterest in using it in robots.

## CHAPTER 10

# ROBOT MAINTENANCE

When a robot malfunctions and a service person must repair the equipment, the task of finding a single fault in the sometimes-very-complicated network of components can be difficult.

## AUTOMATIC DIAGNOSTICS

In many cases, however, the robot can automatically determine the position and nature of the fault. If the robot computer is properly designed, it may allow some memory component to register whatever signal control level instituted a particular stoppage of the robot. If a signal is missing which the computer considers essential to the robot's operation, a message can be put on the operator's display indicating the exact point a repairman could check. When an error in the robot is the fault of the microprocessor component itself, other automatically produced signals can indicate this.

Where robots are used in high-volume production, the saving of even a few moments of diagnostics can save a tremendous value of components that would otherwise not be produced during this diagnostic time (Table 10.1). If rapid diagnostics indicate that some failure will require many hours of service, this can aid the production foreman in his decision making. If a robot motor burns out and this can be determined quickly, the foreman may decide, if it is near the end of the shift, to send the human workers home early. If the same diagnostic process takes an hour, the workers will have remained idle this entire time.

## CAN'T REPAIR ITSELF

Although the robot may be able to diagnose its own faults, it is usually not possible for it to repair itself. The malfunctioning component usually

**Table 10.1. Robot Repairs**

| Rank | Expected Incidents of Malfunction | Expected Time for Repair if Parts Available |
|---|---|---|
| 1 | Operator error | Less than one minute |
| 2 | Program error | Less than one minute |
| 3 | Storage memory error | Several minutes |
| 4 | Connections and cables | One-half hour |
| 5 | Electronic components | Several minutes |
| 6 | Electrical components | One-half hour |
| 7 | Fluid handling components | One hour |
| 8 | Dynamic mechanical | Several hours |
| 9 | Static mechanical | Several hours |

must be replaced before the robot is fixed. If an adjustment must be made, it usually involves the mechanical turning of a screw, a task not possible for the robot. If it were economically feasible to install multiple components of the same type, it is possible the robot computer could switch from a defective component to its spare already installed. However, these additional spare parts would add to the capital expense of the robot; and, if a malfunction did not occur, they would offer no benefit in the operation of the robot.

A particular component also can be damaged by an error in some other signal or process. If the first component were damaged upon automatic switching, the second or spare component is likely to be also. To diagnose this type problem, we could have a more sophisticated robot computer to monitor the input and output signals to each of the components, again adding to the capital cost of the equipment.

It is generally considered more economically prudent to configure the robot, especially in its controller, as a series of modular components by allowing the plug-in replacement of new printed circuit boards and the quick replacement of components such as transformers and motors. The time required for repair can be reduced and the necessity of an automatically adjusted robot eliminated.

While a spare board might cost $1,000, the actual malfunctioning subcomponent of the board may cost only $12. If there are several robots using the same computer board in the same plant, only one or two of each type of board need to be allocated for repair. As these boards were used, a defective board could be examined and inexpensive components replaced on the board, making it functional again.

## DIAGNOSIS BY PHONE

Because robots are increasingly automatic in their diagnostic functions, it is possible to communicate these diagnostics over phone lines. At a particular factory, the service personnel may not be highly trained. If they have access to some far-away expert, they might be capable of performing repairs once the specific needs were established. By connecting the robot's automatic diagnosis through the phone lines to a central repair network, a determination could be made of the robot problem and its potential correction. A set of specific instructions could be printed for a repairman at his location.

Where service must be performed on a 24-hour basis, this allows a few key people to run the night shift of repair guide diagnostics for even a large company. Where computers are used for the storage of the historically documented diagnoses, considerable repair information is available to the local robot repairmen. It is possible for a newcomer to the business to have the experience of many years.

# CHAPTER 11

# JOHN Q. ROBOT GETS A JOB

When an industrial process is being automated with a robot, the work cannot begin immediately after the robot arrives. Unlike the robot's human counterpart, a robot does not walk in the plant doors and, after a brief orientation, begin work. Robots, like other machinery, must be properly installed to accomplish their tasks.

## INSTALLATION

Most robots, especially those of large size and handling capacity, must first be firmly mounted to the floor. The robot must be mounted on a level surface so that elements of its motion can be level with the surrounding equipment. Some manufacturers go so far as to specify an epoxy coating to be applied to the floor before the robot is mounted. Generally, the robot can be fastened directly to the shop floor by anchoring bolts.

If a fine adjustment of the robot's position is necessary, the robot is sometimes mounted to a heavy steel framework which allows the robot to be rigidly mounted yet have its position adjusted. In a pick-and-place robot installation, the exact position of the robot can be essential to its proper operation. The number of positions a pick-and-place robot can reach limits the available positions to which the robot can be mounted. In a pick-and-place robot installation, machinery with which the robot will work often must be reoriented.

In addition to mounting the robot itself, the additional robot equipment must be properly placed. If there is a hydraulic power supply or special cooling equipment, care must be exercised in the placement of these components. The power supplies may not be too distant from the robot itself. The cooling units must be placed conveniently near the hydraulic unit so that their fans may properly cycle cooling air. The robot control cabinet is generally placed far enough from the robot so that no combination of arm positions will allow the robot to strike its own controller. However, the control cabinet must be close enough to the robot to allow the operator a view of the robot's actions.

In some robots, the controls are located conveniently on the robot's body. The operator can stand at the rear of the robot and actuate the control switches, yet be out of the robot's range. In some other units, a hand-held programming module is connected to the main robot controller with a series of wires. The main controller can be located some distance from the robot; yet the programmer, who carries this small module, has access to the robot working place plus the ability to move about while programming.

## PROGRAMMING

Once the robot has been permanently installed in its final position, the program can be written. Sometimes, for expedience, this programming process must begin before the surrounding equipment is actually installed. This presents a problem for the programmer, since the future locations of the surrounding equipment may be unknown. It is possible on most robots, however, to make an approximate program without the peripheral machinery. Then, once the machinery is installed, adjustments can be made for the fine positions of motion.

It is increasingly popular to program robots for one location with a robot physically at another facility. When this is done, a set of robot positions is established in another robot. At this distant location, the robot user creates a program with as many moves and logic sequences as possible. The program can then be transferred to some storage mechanism such as computer tape or an EPROM chip. When the new robot is installed, this premade program can be entered into the computer memory, and the robot is then at least partially programmed.

In some manufacturing operations, the same product is manufactured at several locations or at several points within the same manufacturing facility. Robot programming can be made drastically easier when one program is duplicated many times for other operations. In some large corporations with several plants manufacturing the same parts, a pilot manufacturing line is built. The processes of the appropriate machines and robots are tested and improved until a high degree of quality and reliability is achieved. This pilot plan, or parts of it, are then duplicated in other facilities. All of the physical parameters establishing the relative positions of the machines are documented and repeated at the new plants.

The economic benefits of this system are such that companies will sometimes purchase robots strictly for pilot operation. Once a given task has been proven out, they will begin a new project with the robot, the robot at the pilot plant never actually being engaged in the production of goods. The limitation to a system such as this is in the loss of spontaneous improvement at the local plant level.

## ADJUSTMENTS

Once a robot has been installed and programmed, the manufacturing use of the system can begin. The robot becomes a day-to-day functioning part of the manufacturing process. Sometimes, however, the robot must be adjusted after it has been in operation. In pick-and-place robots, particularly those being used in very high-speed parts transfer, there is a tendency for the mechanical components to wear upon initial use. The exact point of contact on the hard stops may be a sufficiently small area so that even a little wear will cause some deviation in a robot's position. It is good practice on these installations to measure the exact position of the robot arms at installation and then later after the manufacturing process has begun.

Other adjustments that sometimes must be made in pick-and-place robots relate to their ability to control speed. At installation, mechanical adjustments are made to flow control valves, relief valves, and hydraulic shock absorber units. However, a robot may loosen up afterwards, making these adjustments out of tolerance. It is likewise good practice to monitor these adjustment points some time after installation and correct any deviations.

Servo-controlled robots also at times need adjustments. Internal monitoring of such things as power consumption and speed might indicate that the robot is misadjusted. Different conditions at the robot installation will sometimes necessitate adjusting the robot different from the factory adjustment. At times, conditions such as temperature, humidity, voltage level of the main power supply, air supply pressure, air supply contamination, and even atmospheric pressure can necessitate adjustments. Fortunately, it is common practice among robot companies to send a qualified service person to aid in installation. Fine adjustments can then be performed by one specifically trained in the installation of robots.

## COMMUNICATION

As the robot begins its operation with the surrounding equipment, a series of signals must be sent between the functioning machines. At installation, these communication lines are established and the exact meaning of each signal determined. At times, the equipment used with the robot will control the robot. At other times, the robot's excess computing power will be used to control the surrounding machinery. In some cases, both the machinery and the robot will be ultimately controlled by some outside process.

Regardless of the nature of the master control, the communication lines between the robot and surrounding equipment must not fail. If there is some signal, for instance, that allows the robot controller to activate a conveyer, we must be assured this signal will never occur while the robot is actually holding a part on the conveyer. Likewise, if some plant computer is being used to signal the robot that it is time to remove a part from a lathe, we must be completely sure the lathe will not still be turning as the robot grips its part. The cable connections carrying these communication signals must be very reliably attached. It is good practice to run the cables through some impervious protection so that no action of the robot, or accidental use of other machinery, would cause a failure in the lines.

In Chapter 16, a sample communication is given between a robot and two other machines to which it is supplying parts. The machines and the robot are working as independent mechanisms, but the machines are relying on the robot for their start and stop signals. The reliability of the signals has been achieved by the proper selection of the signal type. In each case, a signal to start the machine is achieved by a positive voltage supplied by the robot. A breakage in the line carrying the signal to start would interrupt the flow of current and would not allow the signal to pass. The result of this failure would mean a malfunction in the network but no great harm to the machine or robot.

Conversely, a signal from the robot to stop the machine would be triggered by the interruption of a normally continuous electrical supply. Again, if the line carrying the signal would malfunction, the result would be only a machine not working. If the robot itself should malfunction—if its controller should not be able to indicate a machine stop condition—a human operator could turn off the robot's power supply and hence interrupt the continuous voltage. The machine would receive the same signal as a command to stop from the robot.

Where some action of the machine is necessary before the robot can safely carry out its portion of the overall task, two-way communication is used. For example, when a robot signals a machine, "Open door," it might have to await a signal from the machine indicating the door is open. If the signal necessary for each of these actions is broken, the robot controller would not allow the robot to enter the door opening. Thus, if there is a malfunction in communications, machine time would be lost but the robot would remain safe.

## MASS PRODUCTION

The foundation of America's large-scale production of goods is manufacturing in a continuous process—on an assembly line. That is, the goods stream past a series of workers, each performing some small task on the work piece. This type of production gains its strength from the expertise of the individual workers, the efficient placement of machines, and the readiness of the operators to perform each task.

When a robot is installed as part of a conventional assembly line, a small malfunction stopping the robot can cause great expense to the manufacturer. Because each process in the assembly line receives its work piece from the preceding process, the interruption of work *anywhere* along the line causes an interruption *everywhere* along the line. Indeed, it is common practice for the line to stop when any of its work stations are inoperative.

When there are large numbers of humans working on a conveyer line, these interruptions usually are corrected with great speed. If one worker becomes sick or injured, another worker takes his place almost immediately. The manufacturer sometimes undergoes great expense to keep relief personnel at all times and for all jobs. It is a great irony that people are considered more easily substituted for one another, almost more "mass produced" than are robots.

Sometimes, the hand tools people use on their jobs malfunction and cause an interruption in the work flow. These hand tools usually are inexpensive enough so that several spares can be on hand. And, any interruption is short-lived.

When a robot is taking part in a continuous-process assembly line, the same easy substitutions employed with people and equipment are not available. If the robot malfunctions, it is not possible to replace it with another robot at a moment's notice. Even if the floor mounting would allow the robot to be quickly removed from the line, the current state of technology does not allow one robot to be exactly substituted for another. If two robots were positioned in the same spot and programmed to do exactly the same job, it would be possible, theoretically, to substitute at a moment's notice one machine for another. However, this replacement machine, while not being used, would still cost the manufacturer the same as the machine actually in use.

In some instances, there has been an attempt to program one robot to replace any of a number of robots being used on an assembly line. If 50 robots are being used, the cost to have one additional robot as a substitute would not be too great. However, the technology of modern robot production does not allow the same program to run on both the regular robot and the substitute. Only a few robots manufactured today have the ability to transfer a program to another without some physical adjustment to the second robot. Were one robot to be used as a substitute for any one of a number of robots, the substitute would need to be programmed in each of the 50 work stations. Where a program may be passed from one robot to another, we begin to develop an ability to increase the reliability of robots used on a continuous assembly line.

While robots might seem to be a high-risk addition to an assembly line, there are methods to decrease the risk and expense of a robot malfunction. First, let's consider the expense resulting from some actual malfunction of the robot. In a pneumatically operated pick-and-place robot, there are numerous air seals. Being expendable items, these seals will eventually fail. If an air seal fails on a robot while it is in production, we might estimate it will take a trained robot serviceman 30 minutes to replace it. If the plant has such a serviceman on hand and if he can be reached immediately upon malfunction, we would expect to lose 30 minutes of production time for the entire assembly line at each malfunction.

An assembly line working with great speed might be producing a part in ten seconds. This would convert to 180 parts lost during the 30-minute repair period. What is the value of the parts and how much expense was endured while the line was not moving? If we establish some value of the part in this particular line—say, the assembly of a left-handed whatsis for $4.60—we have lost $828 worth of parts. While we did not expend the raw material for these parts during the 30 minutes, it still was necessary to pay the labor cost and most of the energy cost for their production. The value lost in such a continuous assembly line by frequent robot malfunctions would at times outweigh the benefit of the robot installation. How can we overcome this cost disbenefit?

In places where robot automation has been used for a number of years, we see a subtle change in the method by which parts are transferred in the assembly line. Between adjacent work stations on the line we might find a *surge conveyer*. Or we might see the transfer of the work piece by a *magazine* rather than by a continuous conveyer. A surge conveyer, sometimes called an accumulating conveyer, allows parts from one station of the line to accumulate before they are delivered to the next station of the line. If the accumulation of parts between the two work stations is enough to satisfy the next work station for 30 minutes, a 30-minute robot malfunction should cause no loss of value in the production of the parts. When transfer of the work pieces is accomplished by a magazine system, many production hours' worth of parts may be accumulated between the individual work stations. When a magazine system is used, it is sometimes possible to have an entire day's production accumulated before each work station. This technique allows the robots to be run on a completely unmanned shift. Human workers attend the manufacturing plant during their normal daytime hours, and repairs are made to the robots and other machinery as needed. The workers leave at the end of the day, however, and only the machinery and robots are left to work on the parts. The machinery and robots are programmed so that if there is some malfunction while the plant is not staffed, the equipment can shut itself off. Any malfunction does not stop the entire manufacturing process, only the work cell with the failing equipment.

One of the major detriments to the use of magazine transfer and surge conveyers is the cost of additional work in progress and inventory. In a modern manufacturing facility, work in progress should be minimized so that the capital expense of this work also can be minimized. For each installation, the following should be considered: Which would cause the greater economic harm, an occasional robot or other equipment malfunction resulting in a loss of production, or a certainty of increased cost for work in progress?

## ROBOT SAFETY

When a robot is installed in a manufacturing plant, it must be treated differently from most other machinery. Most machines, by their very nature, are self-contained. That is, they do not operate outside their own physical boundaries. If we do not allow ourselves to get too close to the machine, no harm can come to us.

This is not the case with robots. Robots have extension pieces which reach beyond the bounds of the machine itself. A person standing at what appears to be a safe distance from a working robot would at times be in danger if a malfunction occurs. Because of the popularity of robots and the natural curiosity they arouse, people often gather to watch a robot perform its task. It is good safety practice, then, to eliminate the possibility of human intrusion into the robot's working area.

Robots are often protected by high fences allowing human workers to view the robot's action and service people to watch for any malfunctions, yet not allowing undesirable approach. Robots are also, usually by code, equipped with emergency stop switches allowing people working in the area of the robot to stop it in an emergency. At times, when it is necessary for a worker to be in a robot's area, special safeguards will be constructed. Sometimes, heavy protective beams will be placed on immovable supports. When a person is to enter the robot area, this beam can be positioned and locked in place. The robot program would be written so the robot would not normally approach the beam, but the beam would be fail-safe protection in the event of a malfunction.

It is sometimes suggested that switches be installed at many points on the surface on the robot arms to detect any obstacle or person in the path of the robot's motion and automatically stop the robot. Such devices are possible. But, while such a device might be allowed to protect a machine from damage by a robot, they should not be relied upon to protect human beings. To allow a moving robot access to a human with only the supposed interjection of a signal for protection would be to invite disaster.

Consider another machine of potential danger to humans: a speeding automobile near a street intersection. We do not allow a signal to protect human safety. If a person is crossing the street when a car is approaching at great speed, just because there is an amber light signal indicating the car must stop, we must not assume the pedestrian will be protected if we allow him to walk in front of the racing car.

## ROBOTS ARE DIFFERENT

The new steel-collar workers that have entered the work place in great numbers and in many locations are different from their human counterparts. They must in some ways be pampered in comparison to human beings. They are not easily replaced when they malfunction, and it is sometimes proper to supply them with work in a different way. It is a balance of economic and technological forces that allows robots to be used efficiently in some installations and makes them inappropriate for others.

## THE TEN COMMANDMENTS OF ROBOT SAFETY

1. Robot programs, equipment, and sensors must not be relied upon to protect human safety.
2. While a robot is working, it must be protected from human intrusion into its working area; access doors must be wired into robot controls to prevent all robot action if the doors are opened.
3. Notice of danger to personnel must be in a prominent position on all sides of the robot working area.
4. Emergency-stop buttons, capable of stopping all robot motions and removing all power supplied, must be provided in locations

easily accessed and out of the working range of the robot. All personnel in the area of a robot must be acquainted with its dangers and the use of emergency-stop equipment.

5. Signals and power connections in and out of the robot must not create hazardous situations if signals occur at improper times or are lost during operation.
6. A robot must be programmed, operated, and serviced by trained personnel.
7. If it is necessary for personnel to be within the working range of a robot during programming, great care must be taken that fingers and other body parts are not placed where injury would result in the event of a malfunction. Personnel must never stand in a position where they might be "pinned" if the robot moves without control.
8. A robot operator must know the actions of the robot under his control before they take place. Programs written by another, or stored in some type of robot memory device (tape, computer, storage, etc.), must be documented to allow the operator to use their contents prior to the robot's use.
9. The robot operator must exercise the same care as a human operator using dangerous equipment. Electric cables and hydraulic/pneumatic power lines must be positioned so that operation of the robot and related equipment will not cause breakage or failure. Pressure vessels, flammable liquid tanks, high voltage equipment, etc., must not be brought to risk by robot action. Flame, welding arc, use of high velocity material, etc., must be used in a safe way.
10. Care must be taken that work performed by the robot does not cause undue hazards in the work pieces (sharp edges, sprung members, overheated surfaces, deposits of toxic materials, dust- or oil-covered surfaces causing the pieces to be dropped, etc.).

## CHAPTER 12

# JOB OPENING: VERY DANGEROUS WORK

Robots can be used to accomplish work that poses a danger to human workers. There are three main areas where machines can replicate the actions of humans and, in so doing, reduce the possibility of human injury. These areas are:

1. Dangerous equipment
2. Hazardous environment
3. Dangerous work

## DANGEROUS EQUIPMENT

Machine tools, particularly those involving large-scale manufacturing processes, are often dangerous as a normal function of their work. Machines with rapidly moving parts, parts that are sharp or hot, or saw blades or cutters which might fly apart are dangerous when people are in the paths of these objects. If a human being is to supply or remove parts from such a machine, the potential exists for injury. Although guards may be placed around the moving parts and barriers erected for greater safety, the proximity of human personnel to the working area always allows for some potential danger. Some jobs such as uncoiling rolls of steel, unloading parts that are stacked unsafely, or testing materials to their breaking point (where it is not possible to isolate testing processes) also provide a certain level of jeopardy to the worker. When robots are used in any such jobs, the overall potential for human injury is reduced.

## HAZARDOUS ENVIRONMENT

There are some jobs that must be performed in an environment harmful to humans. One environmental factor whose true danger has recently been recognized is noise. Human exposure to noise has been related to such physical phenomena as hearing loss, unusual mental stress, and effects on basic body functions. The noise levels considered most hazardous to people are those produced by equipment so loud as to be almost beyond the comprehension of those unused to factory work. Machines performing functions such as stamping metal parts, canning, and chipping metal and processes involving high-speed motors can produce sound in very high decibel levels.

As measured on the *Decibel A* (dBA) scale, these machines are considered by current Occupational Safety and Health Administration (OSHA) standards to be unsafe for long-term exposure. Particular machines are rated by their exposure level to the unprotected ear and by the number of hours a person may be legally exposed. If a particular work process requires that no person attend the area longer than two hours per day, the company must change its work force four times in an eight-hour shift. If the work load and the other nonloud hazardous jobs are easily adjusted, this presents no problem. But in some areas, the man-hours lost due to sound exposure can be very high. There is some evidence showing that even acceptable levels of noise exposure can, over a long period, cause permanent hearing loss.

The use of robots in noisy environments can help protect the safety of factory workers. Although at times—while a noisy machine was running—a robot probably would have to be maintained by a person, these times most likely would be short. If some malfunction would occur requiring a long service time, the machine producing the noise would be stopped.

Some work place environments contain material toxic to humans. Particularly in the chemical-processing industry—where material being produced is hazardous to health—there is always some chance of leakage. In most cases, the equipment and machinery used put the potential for human injury so low that it is safe for people to work in the area. In some cases, however, where tanks must be opened, where samples must be withdrawn, where leaks—once encountered—must be repaired, there is a risk factor involved.

There are jobs involving process chemicals such as acid. For example, when automobile radiators are being reconditioned, they must be placed in large tanks. The tanks are filled with an acid solution, and corrosion is chemically removed from the interior of the radiator. The process, even when partially mechanized, requires some human contact with the acid and fumes. The same is true in some instances where sulfuric acid is placed in automotive batteries. For the average auto owner who occasionally checks the acid level in his personal car, the exposure to the acid is very small and very short. But, where this process is carried out in large scale, many hours at a time, the concentration of acid in the environment can be great.

Other hazards to humans are found in the biological industries. Pesticides, even when controlled properly, can expose workers to dangerous chemical substances. The biological creations by such new technology as recombinate *deoxyribonucleic* (DNA) can expose workers to active biological cultures with great potential for human harm.

Some work place environments–contaminated with a toxic substance–are degraded below an acceptable human level by changes in the normal concentration levels of the air. During some processes, the level of oxygen in the air workers breathe can be depleted. Some chemicals in the environment can be safe in low levels but in higher amounts can be very hazardous. Carbon dioxide or nitrogen levels that are too high can be just as damaging as an outright poison. Again, when equipment is handled properly and the process chemicals are safely contained, there is no danger to human workers. But in cases where the likelihood of an accident is great, a robot might be used to protect human safety.

Some toxic elements are considered a danger only with long-term exposure. Currently, a tremendous increase in the study of long-term exposure to asbestos has resulted from statistical information on asbestos workers' health. Other products in the textile industry and in the manufacture and refinement of lead, mercury and other heavy metals–as well as the long-term exposure to wool and some animal products–are stimulating a consideration for the basic safety of workers in these areas. When robots are used for long-term work involving some potentially hazardous material such as asbestos, the machine itself will develop a concentration of the material within its parts. Inside the control cabinet and within the arms of the robot itself will be a buildup of whatever airborne materials the robot has been exposed to. A new potential danger has now been created. Exposure to the robot itself can now be exposure to enhanced concentrations of hazardous materials. Fortunately, since the robot must only be maintained sporadically and since the workers repairing and maintaining the robot would be knowledgeable of the potential hazards, no harm should result.

Some of the most frequent environmental exposures to hazardous conditions are those involving drastic temperatures. As a normal part of the steel-making process, temperatures must be produced far above those encountered in the normal environment. Workers must open doors and work equipment in close proximity to the blast furnace. It has been said that there is no job on the earth that causes such great discomfort as steel making. Once the steel is manufactured, it is again and again reheated. It is heated for forging processes, for hardness treatment, and for forming. The glass and plastics industries also use high temperature as a means to form and purify their products.

Robots are not inherently immune from damage by a high-temperature environment. Many of the semiconductor components used in a robot computer will malfunction at levels humans find safe. Motors and other equipment will sometimes malfunction if applied to temperatures found in steel making and other industries. Hydraulic equipment, although able to withstand temperatures high in human terms, can cause a fire and explosion if placed near an open flame or in high-temperature work places. Hydraulic oil, although not a good combustion source, will nonetheless burn and create smoke.

Robots can be adapted to high-temperature exposure with special equipment. The robot control cabinet and, in some robots, the body of the robot itself can be artificially cooled with air conditioning units. Where the arms of the robot are hollow, cold air can be circulated through the robot, allowing the components to remain cool, even while outside temperatures are very high. In some hydraulic systems, an emulsion of water-soluble oil and water is used in place of the normal hydraulic fluid. This emulsion is not flammable and will allow safer use of the hydraulic equipment. Care must be taken however, because the water in the emulsion—as it encounters rapidly changing pressures—has a tendency to create steam in an enhanced-temperature environment. Malfunctions in the robot can occur from the expansion of the oil by small steam pockets.

Hazards also exist when workers are exposed to low temperatures. In the food-processing industry, workers must sometimes remain for long periods in refrigeration units—for example, where ice cream is manufactured or shipped in bulk. The temperature of a freezer is sometimes zero degrees Fahrenheit. If workers are in good health, dress properly and use caution, exposure to these temperatures should not cause harm. However, long-term exposure can lead to chronic health problems. Where the work performed requires a high level of physical effort by the worker, an increased breathing rate causes the lungs to be exposed to very low temperature conditions.

Robots function well in low temperatures, but some adjustments may be necessary for proper functioning. For example, lubrication on the mechanical components may be changed to allow lower viscosity at lower temperature, but the components themselves adapt well to a low-temperature environment.

## PERILOUS WORK

There is some work that is intrinsically dangerous; for example, the manufacture of explosives, the testing of firearms, the use of lasers and beamed energy equipment, and the industrial use of X-rays, radiation, and very high voltages. There is no place involving these intrinsically dangerous processes where a human can work with complete safety. Ideally, such processes would be completely automated and beyond the need for a human worker. As a practical matter, however, there are workers daily engaged in dangerous activity. With the new interest in robots for industry, some workers and their union representation have insisted that robots be employed in hazardous processes. The hope is that robots can perform all perilous tasks with a human directing the actions from a position of complete safety.

# CHAPTER 13

# UNTOUCHED BY HUMAN HANDS

An environmental consideration in the use of robots is potential contamination of work in process by human workers. Some areas in which this contamination would normally be considered a problem are the food industry, the medical profession and some research and testing.

## CONTAMINATION

The food industry uses a great many manufacturing processes to purify and to sterilize the food we eat. But at times, contamination can be introduced into the food by human workers. Robots could be used for some processes. They are intrinsically clean in the way they operate. If there is some transfer of robot medium–such as paint from the robot arm or dripping lubrication–into the work piece, this phenomenon can quickly be identified and prevented. Further, a robot can be designed so that it will be possible to clean it repeatedly without harm to the components. Cleaning solutions both caustic and abrasive can be used on robots while not generally on humans.

In medical applications where the isolation of a patient is necessary, robots could be used to provide services sometimes carried out by a nurse. Although, with current technology, a robot should not be considered automatic enough to give injections or back rubs, it could nonetheless reduce the contact of the patient to the outside world. A robot could be used to deliver medication that a patient could administer to himself. Robots could be used to deliver food, to carry samples, even to make beds. Where potentially dangerous medical material is to be transported or processed, the robot could perform the work without contact to human tissue.

There is a potential use for robots in medical research. For example, when virus to be examined for bacteria must be carefully isolated. It is possible for a research worker to quite innocently contaminate a sample without knowing it. A great deal of time could be lost to the researcher because of this contamination. In testing, erroneous results might be received because of the proximity of the research work to a human. The use of robots could allow samples to be conditioned and research experiments to be prepared without risk of contamination. Certainly, there are now safeguards in the form of automatic equipment and proper test procedures, but the use of robots would add another factor of safety.

## GRIPPERS: THE HANDS OF A ROBOT

Much like our human arm, the arms of a robot are not very useful without a hand and grasping fingers. The devices mounted to the working end of a robot arm are generically referred to as grippers. Although the term *end-of-arm tooling* is used to describe virtually all the devices which might be mounted on the end of the robot arm, the term *grippers* is used only for those devices which will grasp, carry, and release some nonrobot component (Figure 13.1 a-f).

Because the tasks assigned to robots vary so greatly, the gripper fingers have a greater variety than any other robot component. A manufacturer that produces one or two robot models might have ten versions of common grippers. Furthermore, most grippers are made with easily replaceable figures. The gripper fingers are often custom made for the individual job at hand. There are literally tens of thousands of unique designs of robot fingers in use today. Although most robots are used in some process type work not necessitating their grasping an object (spot welding, arc welding, spray painting), a large percentage of robots is engaged in the task of moving products.

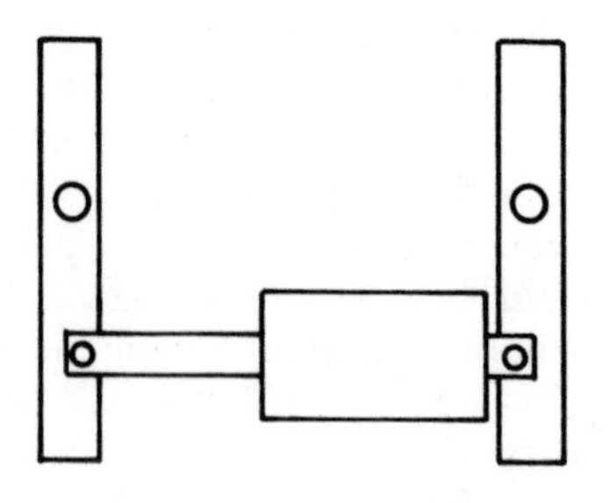

**Figure 13.1-a.** A single side-motion gripper.

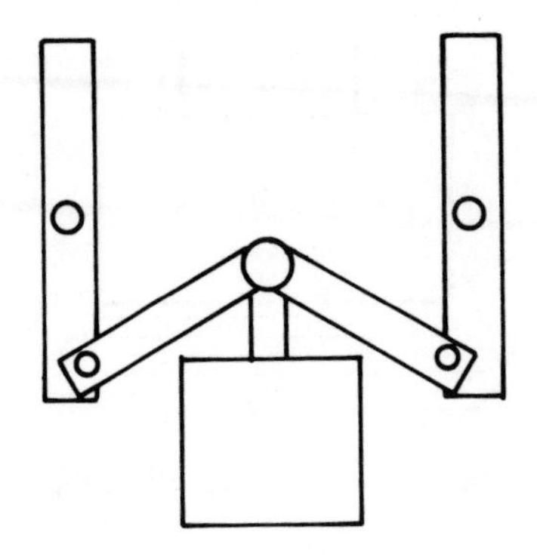

**Figure 13.1-b.** A linkage-type gripper.

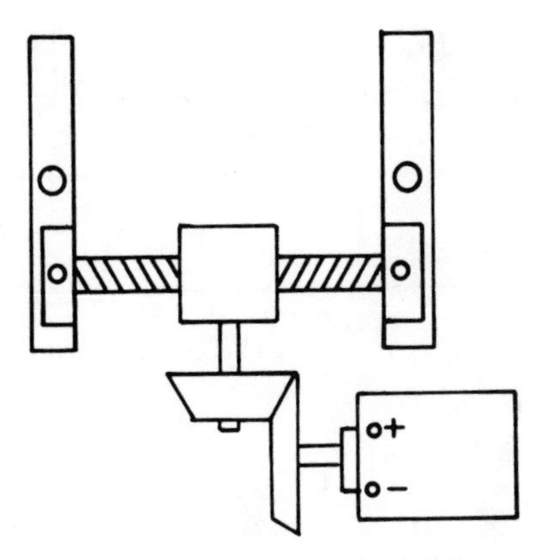


**Figure 13.1-c.** An electrically driven gripper.

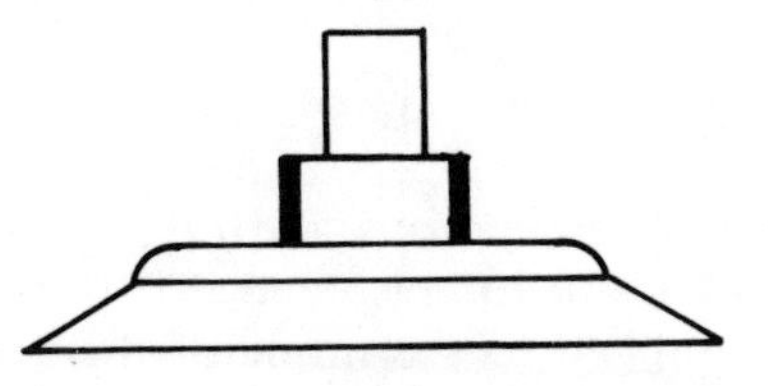

**Figure 13.1-d.** A vacuum gripper.

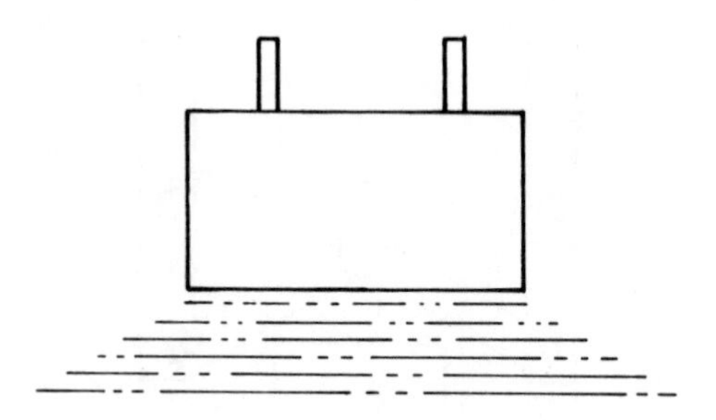

**Figure 13.1-e.** A magnetic gripper.

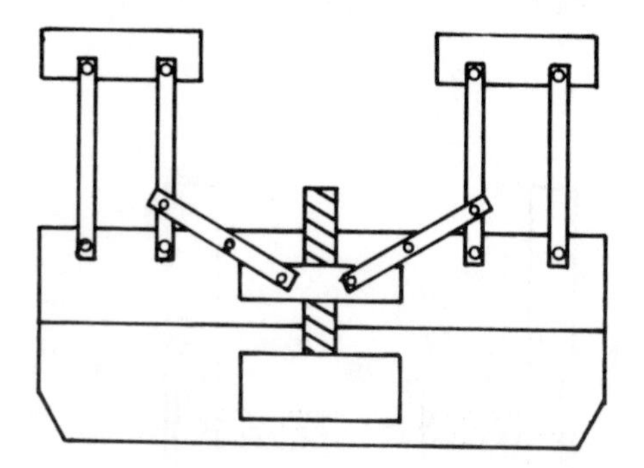

**Figure 13.1-f.** A parallel-motion gripper.

The simplest type of gripper is some motion-producing device joined to two fingers. The fingers may be open or closed and they may grasp some object. Various improvements on this design can be made, but the central purpose and motion of the fingers remain the same. Some of the versions commonly used are mechanical linkages to hold the object in grip even if the power supply is moved, to keep the fingers parallel while they are closing, or to provide a sensitive or variable source while gripping.

If the part to be grasped has some special properties about it, we may make use of other types of grippers that do not approximate human form. If the part is both nonporous and relatively flat, we may make use of a simple vacuum cup to grasp and lift it. To release this object, a blast of air can be applied to the inside of the vacuum cup; it is said that we *blow off* the part. If the object to be grasped has some magnetic property, it is possible to use an electromagnet as our gripper. To release the part, we simply change the magnetic field electrically.

When there is a series of relatively small parts to be grasped at the same time, a single gripper with multiple fingers can be used. Figure 13.2 shows such an arrangement for placing small parts in a line. If the parts to be picked up have some particular surface feature that allows us to grasp them, we may make use of a specially designed gripper.

Figure 13.3 shows a special gripper for picking up bottles. As the gripper is inserted in the neck of the bottle and an air cylinder actuated, a rubber section of the gripper expands much like a thermos bottle seal to grip the bottle. Similarly, it is possible to make a doughnut-shaped gripper (Figure 13.4) out of flexible material. It's placed over the outside of a bottle. The gripper's interior sector expands, and the outside of the bottle is grasped. These last two types of gripper are particularly useful in handling something like glass, because the only contact between the gripper and the glass is the compliant material.

Also in the realm of the expanding set of fingers is the one shown by Figure 13.5. Here, the portion of the object to be grasped is within a cavity of the object, and it is necessary to have fingers which will expand in the slot. The expanding fingers must be made to fit the object exactly to avoid interference as they enter the part and to allow

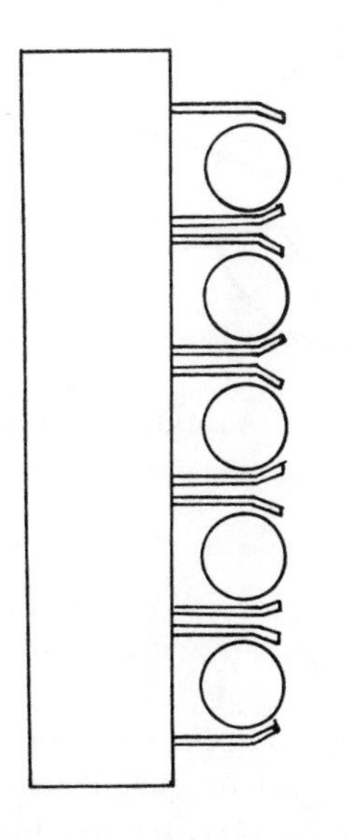

**Figure 13.2.** A multiple-finger gripper.

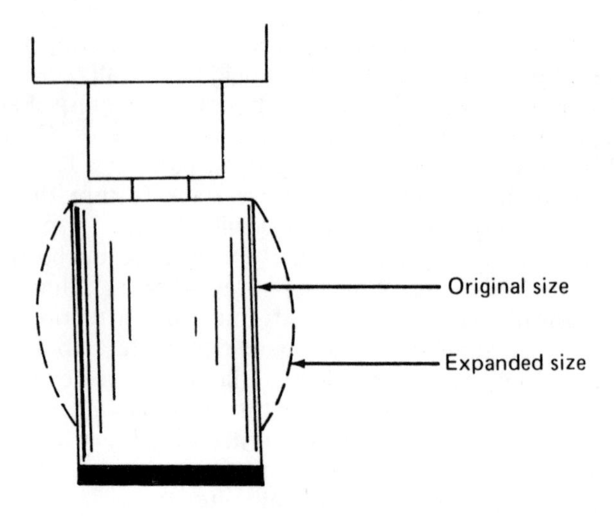


**Figure 13.3.** An expanding rubber-washer gripper.

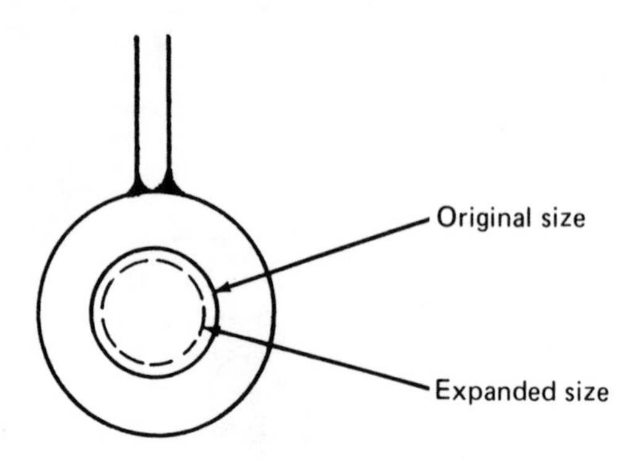


**Figure 13.4.** A shrinking-doughnut gripper.

accurate gripping, once in place. When the object to be gripped is very heavy, or when it has a type of surface which would require great force to grip, it is possible to use a gripping mechanism with very high force output. This is usually accomplished using hydraulic pressure. If the object to be gripped is steel or cast iron—so that it might withstand sufficient pressure—and has very robust features of its own, weights of hundreds of pounds can be moved with great ease.

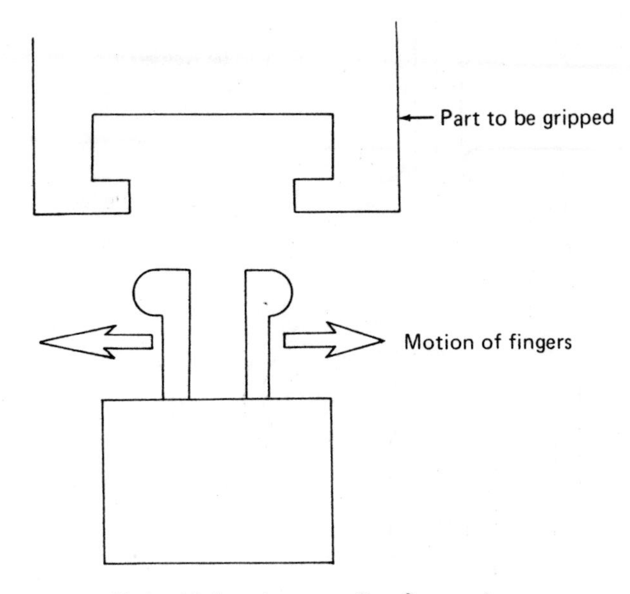


**Figure 13.5.** An expanding-finger gripper.

Some of the other components added to the end of a robot arm, although not to be considered as grippers, are shown in Figures 13.6 through 13.8.

Spot welding guns are often used with robots, because of the relative weight and handling awkwardness of the gun itself. Because the jaws of a spot welding unit are so heavy—as they must be connected to water cooling lines, air pressure lines (to facilitate their gripping power), and high current wiring—they are difficult to maneuver manually. Furthermore, when the spot welding operation must take place in confined space—under the hood of a car, for instance—it is difficult to provide an overhead type of support to carry the weight of the spot welding gun. Robots, not feeling the weaknesses of human fatigue, can work long hours moving the heavy spot welding guns and performing the welding task.

Arc welding guns are, on the other hand, very light and easy to move. They are often placed on the arms of robots, not because of their weight, but because of the ability of a robot to duplicate a task many times. When arc welding is performed manually, great precautions must be taken to protect the human welder from harm. Heavy clothes must be worn, special insulated gloves used, and a special shield put on the face allowing the welding operator to see through heavily tinted glass. When

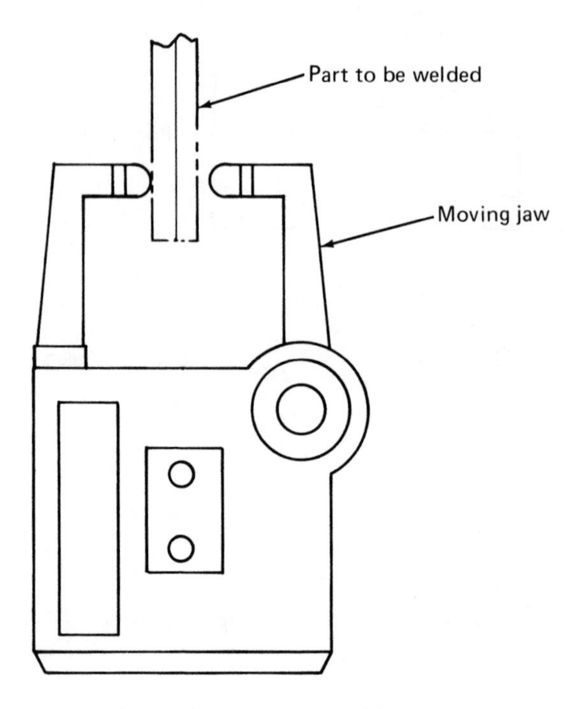


**Figure 13.6.** A spot-welding gun.

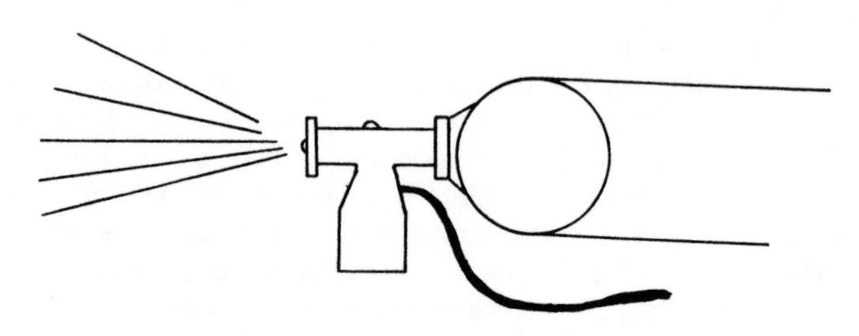

**Figure 13.7.** A spray-painting gun.

great precision is required in the weld, these precautions are a disadvantage.

The robot welder is not encumbered by protective garments. It may be programmed to follow the path needed for a weld while the welding equipment is not in operation. Welding operations formerly performed

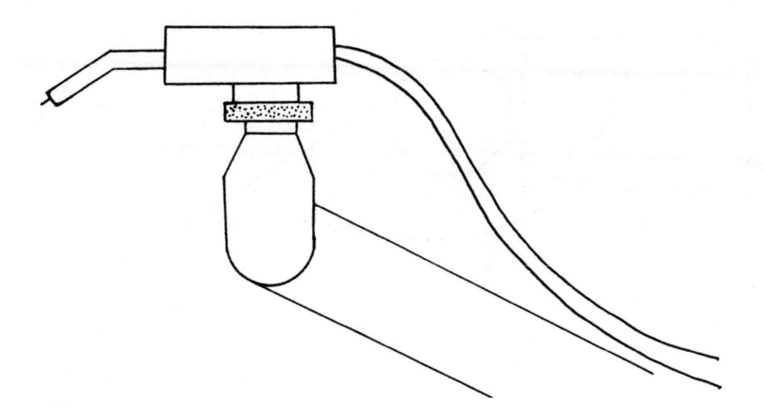

**Figure 13.8.** An arc-welding gun.

over a time span adjusted to the ability of the human operator can now be sped up to the ability of the robot. Although no person might be able to perform the task in real time, the robot, once programmed to do so, can accomplish the task.

Deburring and camera manipulation are also suitable robot functions (Figures 13.9 and 13.10).

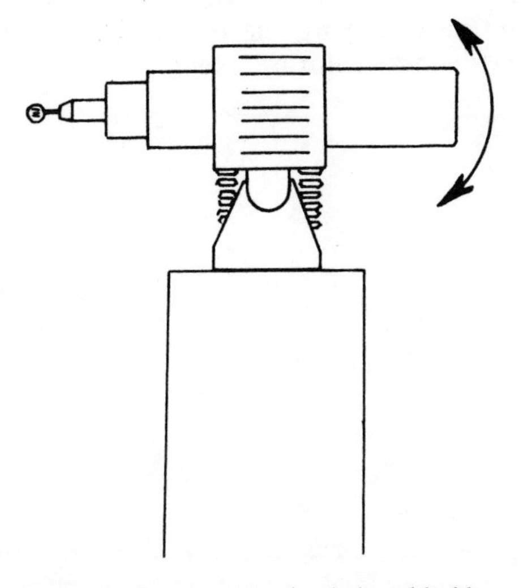

**Figure 13.9.** A spring-loaded tool holder.

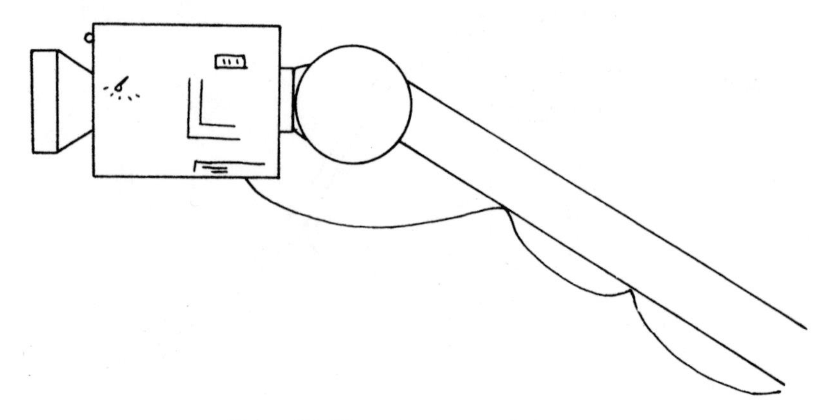

**Figure 13.10.** A robot-carried camera.

The spray-painting function is provided by only some of the robot manufacturers. This is not because the spray-painting task is difficult, just the opposite. The task is among the easiest for a robot and may be performed by the least expensive robot varieties. Because of price competition and low profit potentials many robot companies do not quote robot spray-painting equipment at all. Robot equipment dedicated to the spray-painting task is often of the tracking computer type. The programming used amounts to the performance of the task in real time and its subsequent repeating by the robot. The accuracy and control of speed necessary for spray painting is low.

# CHAPTER 14

## SERVING WITH NO EYES TO SEE

Robots are capable of doing some jobs that human beings are not. It is possible for the robot programmer to conceive of a task which can be accomplished by the robot but of which the operator himself is incapable. An example is the welding of two pipes. If two pipes must be joined as in Figure 14.1, it would be a fairly simple task for a human welder to weld the pipes along the regular seam where they join. If the weld must be placed on the inside of the seam, the task would become more difficult. If the pipe is not large enough to allow the welder to enter the pipe, the weld must be performed while he is standing outside. Although the welder may be capable of putting his arm in the pipe and physically reaching the weld points, without his ability to see the weld as it progresses, it is not likely that a good weld could be accomplished.

If an available robot is capable of putting an arm in the pipe and physically able to reach all of the points, it could be programmed to do this welding operation. The program must be established by a human but could be done so with great ease. The first part of the program might be written to allow the robot arm to enter the pipe without touching the sides. Once this portion of the program was written, the robot arm could be withdrawn from the pipe and one section of the pipe removed from the working area. The written portion could then again be executed, placing the robot in proximity to the weld along its path of safe entry.

With the one pipe removed, the operator would now see the path the robot welder would follow and correctly identify the points in the weld path to be part of the program. Once this portion of the program was completed, the robot could again be withdrawn from the work site. The original second section of pipe could now be returned to its preweld position and the robot program continued. The robot would enter the pipe and perform its motions of welding. Now, the programmer

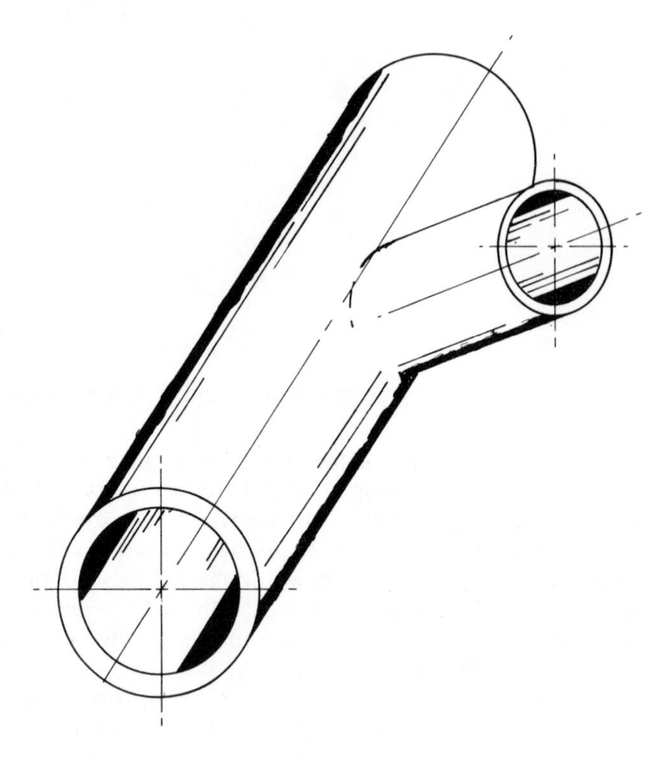

**Figure 14.1.** Two pipes to be welded by a robot.

could write the final portion of the program to allow the robot to withdraw from the pipe without touching the sides. This program, once written, could be run as often as necessary to weld two similar pipes.

Jobs such as this increase the human ability to build and create. Manufacturing need not then be limited only to what we are able to do ourselves or what we might do with special-purpose machinery. It is limited only by the bounds of our imagination and the ability to conceive the potential for robots.

If it is possible to create, at our own pace, robot actions which will be performed with greater speed, to correct many times the same operation until it is perfect, and to allow the robot to execute a task with perfection without ever once doing it ourselves, it is possible to have tasks performed considered until now unrepeatable. It might be possible to duplicate the Mona Lisa by the use of a robot. As a practical matter, this will most likely not be done, because the thousands of brush strokes made with extremely fine motions would require perhaps years of continuous trial and error by the robot programmer.

The technique used to make such a copy, however, is also available to us for new creations. A robot could be used to trace a human face while a person was sitting in a chair. Once the feedback system of the robot received all the topography information of the face, the robot could be used to carve a duplicate. With changes made in the basic information by the robot's computer, other versions of the carved face could be created. The face could be enlarged, the face could be altered, the face could be distorted. An individual artist's ability to create would then not be hampered by his ability as a craftsman to physically perform the tasks he desires.

Although the use of a robot as an artist's tool may be limited, similar techniques can be used to create custom products to exactly match some need of the customer; for example, silverware etched with a family crest or symbol.

In addition to being able to accomplish some tasks that humans cannot, robots can work under circumstances that people would not tolerate. If an area of a plant is operated completely by robots, the lights could be turned off, except for human inspection. There would also be:

- no bathrooms
- no cafeteria
- no parking lot
- no theft
- no lawsuits
- no additional personnel department
- no additional payroll department
- no breaks
- no vacations
- no social security

As for negative features of robots, robots cannot be laid off. A robot costs almost as much when it is not running as when it is. Robots cannot be promoted, buy company stock or contribute to suggestion programs.

*The workers were dutiful in every way—*
*Able, proficient, at work every day.*
*Managers were proud of their robot work force.*
*Stockholders were happy with high profits, of course.*
*The flow of good products was never disrupted.*
*Inhuman workers will do whatever instructed,*
*Happy to help when reprogrammed one night*
*To load the thieves' truck, then burn down the site.*

# CHAPTER 15

## WHY DON'T ROBOTS HAVE ALL THE JOBS?

By some accounts in the robotic business, robots are capable of doing almost any job. They are superior to their human counterparts in ability and speed. Why, then, haven't the robots taken a greater share of the labor market?

To answer this question, we should look at why robots are used in the first place. The basic reason for the use of robots is economics. A manufacturer uses robots to produce his goods at a lower price and at a higher quality for a given investment. Robots, unlike human workers, do not tire from their work. They produce the same quality goods after a million parts as they do after a thousand. Because robots have very uniform production, it is possible to exactly determine the cost of a product before it is produced. A very low margin may be maintained, helping to make the manufacturer competitive.

Robots offer this service, however, at a premium in capital investment. The capital needed initially to place a worker in a factory is much lower than the comparative start-up cost of a robot doing the same job. A human worker needs tools and must be trained, and often these costs must be considered as capital investment. But, overall, the robot requires more money before the goods are produced if the comparative labor is paid for as it is used. If the worker works for one week, he is paid for one week. If the worker is not needed, he will be laid off and therefore of no cost to the company. If a robot is producing goods and those goods are not in demand, unless the machine is sold the manufacturer must still maintain his original capital investment.

A modern industrial robot that will accomplish tasks using servo-type control with a computer memory has a price to the manufacturer on the order of $50,000. There are some robots—for doing either very large or very complex tasks—which can cost $100,000 to $150,000. This is for the bare robot; no tooling, no special fixturing, no automatic machinery to supply the robot with its parts.

An installation using a robot doing even a simple task can cost $125,000. This would be a task that perhaps four people could do manually. When we look at these costs, it seems foolhardy to use robots at all, until we analyze the amount of goods that can be produced by both the human workers and the robot. The robot's overall production cycle can be so much faster than its human counterpart that the goods produced can have a drastically lower price.

A typical unit of measurement used to calculate the benefit of a robot is the robot's *return on investment* (ROI) or pay-back. How long does it take the robot, when producing goods, to accumulate enough extra profit so that the robot will pay for itself? This is no different than the computation used for any other type of automatic equipment. A typical robot in industry today can be expected to have a pay-back on the order of two and one-half years. There are instances in industry where the job accomplished by the robot is very complex and the parts worked on very expensive, whereby the pay-back for the robot is computed to be in a few months.

At current relatively high interest rates, it is difficult to justify robot application if the pay-back is longer than a period of, say, five years. This is artificially determined because there is industrial equipment—even entire plants—in use with an ROI of longer than ten years. Something to remember, though, is the rapid change in the robot industry. A robot purchased today may be obsolete in five years. Certainly, much of the equipment purchased for robotic applications five years ago is no longer in use. It is not that robots have built-in obsolescence or that they do not have a long useful life. Indeed, robots are so well made, on the average, that they will last ten or 15 years in continuous use. Rather, five years hence, the robot equipment available for a certain dollar investment may be so superior to that produced today that it would be beneficial to accept a loss on the existing equipment and purchase the new equipment.

Another factor to consider when computing the respective number of robots that will replace human workers is the ability of the robot to replace hard automation. When a manufacturer contemplates automating a particular task, there are currently three general categories to be considered. The first is the human-assisted machine; that is, some type of automatic machine allowing a human operator to do some of the work. The second category is a robot which works in combination with a general purpose machine. The third category is a completely automatic system wherein the machine itself is specifically designed to do the job at hand—hard automation.

If, for some economic or quality reason, it is decided that a general purpose tool coupled with manual labor will not be adequate to do a particular job, it is possible that a robot plus manual labor can be used. So, even with robots, there is still a place for human workers.

## IS A ROBOT GOING TO TAKE YOUR JOB?

Until now, we have been talking about the economics of robots in a very abstract way. It is possible to consider a certain amount of displacement in the work force due to robots and still keep our emotional perspective. However, when we confront the issue "Are we to be replaced by a robot?", the matter becomes more personalized.

The decision of whether a robot will take our job is based on what we do in the work field. If the task we perform is one easily done by a robot and if we are paid a substantial wage, it is likely our job is a candidate for robot automation. One of the major reasons for the use of robots by the auto industry is not particularly what is done in automobile manufacturing, nor the type of product produced. Rather it is that most auto manufacturing fits the criteria of easy robot use and high worker pay.

In some other industries, the work is repetitive and of sufficient ease that a robot could handle the job. However, the workers are paid low wages. So it is more economical for a company to continue using manual labor than making the capital investment for robots.

An example is the textile industry. There is little work-function difference between stitching together cloth for a coat and stitching together cloth or vinyl for an auto seat. However, robots are used in the production of auto seats and not in the production of coats. There are some additional reasons why robots are or are not used. One is the size of the company. A small garment district tailoring operation is on a totally different scale than, for instance, General Motors. The tailoring company cannot afford the capital expenditures of the auto giant. Another reason is the degree of precision needed for the individual job. If a coat is made slightly too small, it is merely sold as a different size. When an auto seat is made too small, it cannot fit a slightly smaller car (size 9 petite). Still another reason for using or not using a robot is the time factor. Auto seats for a particular automobile may involve years of planning while a coat is designed one week, manufactured the next, and sold in the stores practically before the sewing machine has stopped. By now, this analysis should have caused great fear for anyone employed in the auto industry, but take heart. Robots will not replace all of the jobs even if they meet the criteria.

There are yet other factors to be considered in large-scale use of robots. If some large automotive manufacturer were, tomorrow, to completely automate its entire manufacturing process and every worker producing goods were replaced by a robot, there would still be a great percentage of those same people working in the factories. Some jobs may have been dissolved, but people formerly performing those jobs would for the most part remain as part of the operation–repairing, lubricating and overseeing the robots.

An analogy of this situation can be found in the computer industry. It was thought with the first use of computers that they would replace a great many office and clerical workers. It was thought that once there was a computer to generate files and sort information, less time would be spent by secretaries doing this work. It was assumed, once a computer could operate a printing machine and generate letter-perfect copies of letters, that typists would be almost unnecessary.

When banks began to use computers, people in the accounting departments assumed that, although there would be jobs for a few specialized computer programmers, the number of accounting personnel would go down.

History indicates that although there were many changes in the way these jobs were performed–certainly the day-to-day routine of sorting files and entering numbers–the absolute number of people employed in the word processing and number processing industries increased. The wages of those people in some cases increased dramatically.

What we expect to happen in the robot industry is similar to what happened in the computer industry. The number of people required to build a robot can be estimated as a factor of the cost of the robot. The amount of labor in a particular robot is only a part of the overall manufacturing costs, the other costs being in the machinery to produce the components and in the components and raw materials purchased on the outside. But if we look at the outside-purchased tools and supplies, we see each was produced with the help of human labor.

The very nature of economics dictates that less actual labor is encompassed in a parts-producing machine than would be in the parts produced manually. But since all of the labor must be supplied to build the machine before it is used for even the first part, we can see, at least on a short-term basis, a net positive labor effect in the use of automatic machinery.

The ultimate cost of manufactured goods produced with the help of robots can counterbalance economic forces such as inflation. While the cost of raw materials is rising, the cost of the manufacturing effort to turn those raw materials into finished products can drop. The net effect can be a long-term lowering of the price of finished goods.

Although no economy can stand the stress of an immediate change in its entire work force, the introduction of robots into western technology over the next decade can provide unprecedented economic return. Since the beginning of the industrial revolution, there has been no single technology with such far-reaching potential as the use of robots and computers. Computers primarily affect the information, entertainment and administration of our society, but it is robots that have the greatest potential for changing the physical world and the way we approach our daily work lives.

To the individual worker whose job is displaced by a robot, this is not a comforting fact. But for the economy in general, we see a short-term stimulation of the job market. Indeed, when a manufacturer enters the robot market to make its own robots for its own use, there is–rather than a replacement of human workers–simply a change from the manufacturing of goods to the manufacturing of machinery to make the goods.

It is too simple to say that when a robot starts performing a job formerly done by people, the people are no longer necessary. The current state of the art in robot manufacture does not allow robots to be completely unattended. Routine maintenance must be done on the robots. As they malfunction and break down, they must be repaired. And because of the high cost of the robots and the value of the components being produced, robot repairmen must be close at hand so that any breakdown can be remedied with little loss of manufacturing.

Depending on the length of the cycle that a robot's job will take, it is necessary to reprogram the robot. With training, it is possible that the robot programmer will be the same person that the robot replaced.

Let's say that a very sophisticated welding robot has been installed in a particular location replacing a worker who had been on the job many years. It is possible that the engineer who installed the robot is a very capable programmer, but not an adequate welder.

As the robot performs its tasks, it may malfunction because of a peculiar situation involving changes in the material or the welding dynamics. The robot programmer may be at a loss to make the changes in the programming necessary to carry out the robot task. The displaced worker, however, may be an ideal person to assess the situation and make necessary changes in the robot program. This would be an example of a human being being used to his full potential. That is, he would be using his experience, judgment and values whereas the robot would be performing the tedious, demanding physical labor that was formerly a part of the welder's job.

Of course, this example is predicted on some assumptions. First, we assume the worker is allowed to get the training to do the programming. Second, we assume that the welding robot is a type sufficiently easy to program. And third, we assume the worker will be capable of

Table 15.1 Robot Jobs

| Jobs | Approximate Training Needed | Number Needed/Robot |
|---|---|---|
| Robot Operator | Few days on job | 1/6 |
| Robot Designer | A four-year college degree | 1/50 |
| Application Designer | Two- or four-year degree | 1/25 |
| Robot Programmer | Several weeks of training special to robots | 1/20 |
| Robot Maintenance | Specific technical background | 1/8 |
| Robot Research | Four-year degree, advanced degree helpful | 1/100 |

understanding programming. Programming work, although quite different from some manual task such as welding, is well within the ability of most workers to understand.

Once again, an individual who sees his job easily and economically undertaken by a robot can be justifiably fearful. But with a proper attitude toward the new careers emerging in the robot industry, he can look to a brighter future (Table 15.1).

## AN ECONOMIC EXAMPLE

As a way to demonstrate the comparative economic advantages of the different types of manufacturing, we will examine an imaginary situation where work might be accomplished by three different methods. The numbers used should not be considered representative of the real world (although the calculations are the type to be expected). Rather, this example will illustrate the interplay of economic forces as they relate to the choice of a manufacturing method. Human, robot, and hard-automation production methods will be considered for the manufacturing of a common part. For each, we will analyze the cost factors and potential profits.

Table 15.2 shows the significant cost factors of the three manufacturing methods.

Our first calculation is to determine the cost to manufacture a single part. We will compute the cost based on a year's production, for one full unit of production (we might be able to hire a worker for four hours-per-day, but we cannot buy half a robot). The formula is:

**Table 15.2 Costs of Three Manufacturing Methods (Per Year)**

| | Human Expense | Robot Expense | Hard Automation Expense |
|---|---|---|---|
| Capital Investment Per Year[1] | $4,000 | $29,257[2] | $52,195[3] |
| Number of Parts | 11/hr = 22,000 | 23/hr = 46,000 | 45/hr = 90,000 |
| Number of Parts/ Cost Rejected[4] | 800/$1.69 | 50/$2.04 | 50/$2.17 |
| Cost of Maintenance[5] | $600 | 1/8 man @ total $26,000 3,000 parts | 1/12 man @ $26,000 2,000 parts |
| Cost of Production[6]/Year[7] | $450 energy 900 sq/ft = $270 $24,000 (direct labor) | $800 energy 750 sq/ft = $225 1/6 @ $24,000 total | $600 energy 600 sq/ft = $180 1/8 @ $24,000 total |
| Cost of Raw Material/Part[8] | $2.60 | $2.95 | $3.08 |
| Value Factor[9] | +1% | 0 | +1/2% |
| Time Necessary For Minor Change | 1 hour | 1 day | 1 month |
| Cost Minor Change | $400 | $2,000 | $12,000 |
| Insurance Cost | 9¢/part | 1¢/part | 1¢/part |
| Salvage | 91¢ | 91¢ | 91¢ |

1. Capital investment includes engineering, laboratory, maintenance tools, interest, and machine cost, divided by life expectancy
2. Robot @ $100,000 + $28,000 engineering @ 15% for calculated use of 7 years
3. Hard automation $225,000 + $61,000 @ 15% for calculated use of 10 years
4. Material cost + disposal cost − salvage
5. Includes special tools
6. Cost of production energy/floor space/man-hours
7. 30¢/square foot/year
8. More expensive raw materials needed for machine production; premachining of part needed for hard automation
9. Value factor = increased value of part made by this means; 1/(value factor) equals the cost factor

$$\frac{\text{Capital + Maintenance + Production + Reject Costs}}{\text{Number of Parts}} + \text{Raw Materials}$$

$$+ \text{Insurance Costs} \times \text{Value Factor} = \text{Cost Per Part}$$

For the human workers, we have:

$$\frac{4{,}000 + 600 + 24{,}720 + 1{,}352}{22{,}000} = 1.39 + 2.60 + .09 \times .99 = \$4.04$$

For the robot, the cost is:

$$\frac{29{,}257 + 6{,}250 + 5{,}025 + 102}{46{,}000} = .88 + 2.95 + .01 \times 1.00 = \$3.84$$

The hard automation cost is:

$$\frac{52{,}195 + 4{,}167 + 3{,}780 + 109}{90{,}000} = .67 + 3.08 + .01 \times .995 = \$3.74$$

If our customers will buy 88,000 parts this year, we would have the option of using any of the three types of production. Four human workers = two robots = one dedicated machine.

If the market for our parts will remain stable over the useful life of the production equipment, the least expensive way to produce our parts is by hard automation. The company may have other factors to weigh in its decision. There may not be time to design and build the dedicated machine, there may be a genuine interest in the employment of long-time employees, or the company may not be able to get the credit it needs to purchase the equipment. Based on the information at hand, however, the lowest-cost method is hard automation.

The cost factors are different if our market changes. First, let's consider the costs if our output is increased to 176,000 units per year. Running a second shift with the human workers would entail an additional man-hour cost of 10 percent. Since everything else would stay the same, the average cost-per-part would be $4.09.

For the robot, the computation for the additional production would be:

$$\frac{6{,}600 + 5{,}425 + 102}{46{,}000} = .26 + 2.95 + .01 \times 1.00$$

$$= 3.22 \quad \frac{3.22 + 3.84}{2} = \$3.53 \text{ average}$$

Additional product produced on the hard automation would be:

$$\frac{217 + 4{,}167 + 4{,}080}{90{,}000} = .09 + 3.08 + .01 \times .995$$

$$= 3.16 \quad \frac{3.16 + 3.74}{2} = \$3.45 \text{ average}$$

With the increased demand, it would still be more economical to use hard automation as our means of production.

Now, let's consider the costs if our product demand fell by 50 percent.

The human work force would have an additional incumbrance of \$8,000 for the equipment and \$540 for floor space that would have to be amortized over the smaller number of parts. This would increase the cost of the parts actually produced to \$4.23 (19¢ more).

One full robot would have to be amortized over the smaller number of parts, along with the floor space that could not be used for some other project. The computation would be: (29,257 + 225)/46,000 = .64 per part extra. The new cost of the robot part would be \$4.48.

Since there is only one machine in the hard automation production, we would have to recalculate the cost for the lower production entirely:

$$\frac{52{,}195 + 2{,}048 + 1{,}980 + 55}{46{,}000} = 1.22 + 3.08 + .02 \times .995 = \$4.30 \text{ per part}$$

It now would cost less to manufacture the part with human labor than with either of the other two methods.

Perhaps the market for our product is cyclic, and plans to achieve a return on investment for the hard automation will be overturned. The choice for either of the capital-intensive forms of production represents a risk by the company's management.

Another part of the risk of capital-intensive production is the chance for change in the design of the part. If, after just a few months of production, the market changes and to sell any of our parts we must redesign them, the cost factors would be as follows (Table 15.3):

The cost to retrain the human workers and the cost for new tools would be \$1,800 or two cents per part.

The cost for re-engineering the robot and the total retrofitting would be \$5,000, averaged over seven years, or three cents per part.

A completely new machine, for hard automation production, would cost at least as much as the old one. If the cost of the new machine were averaged over the next ten years, the cost per part would be six cents. But if the amount were not averaged, the direct loss to the

**Table 15.3 Loss Due to Change in Product**

| Manufacturing Method | Dollar Loss | Loss Per Part |
|---|---|---|
| Human | 1,800 | 2¢ |
| Robot | 5,000 | 3¢ |
| Hard Automation | 50,000 | 6¢ |

company would be $50,000. Although some advantages are seen for each type of production in the above examples, it would be a tough decision to choose the best one for a particular company.

In the real world, some of the factors considered are far less balanced and make the decision clear. Using the data provided, you may wish to calculate for yourself the relative merits of each type of production under the following changes:

- A change in the design of the part every year
- A value factor for the parts made by the robot of +15%
- An increased labor rate of 15%
- A manual part rejection rate of 15%
- A change to three-shift production
- A fire destroying the machines (A new robot would take only three weeks to install.)

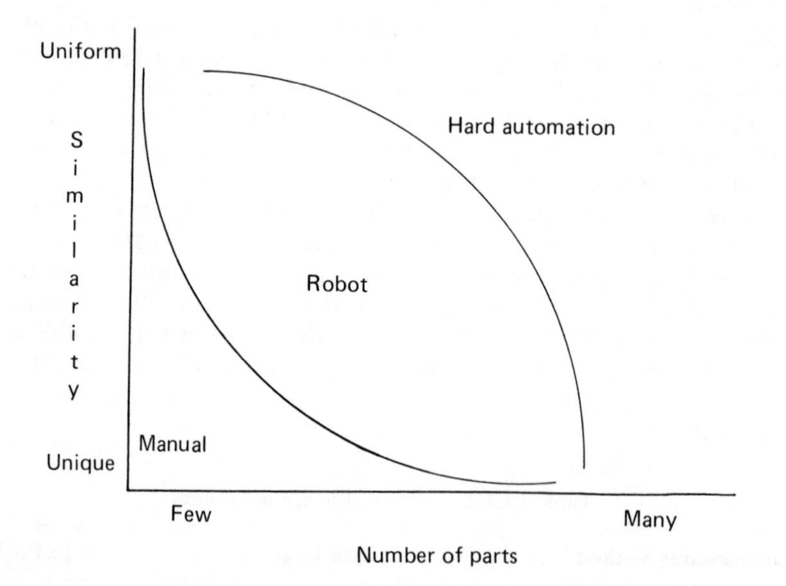


**Figure 15.1.** The general areas for the profitable use of robots and hard automation.

- A down-time factor for the machines of 15% (What's the value of the maintenance?)

Figure 15.1 shows the general trend of profitable production for each of the three types of production.

## ROBOTS AS A PREMANUFACTURED SOURCE OF HUMAN LABOR

As we examined our model, it appeared that the human labor component in the manufacturing cost of a part was greatly reduced with the use of robots and hard automation. We must consider, however, that each of the machines used in place of human labor also represents the work of the people that originally made the machine. Each of the parts used by the machine manufacturer, even if purchased outside the company, represents human labor in its manufacturing. In turn, each of these components, when produced in some part by machinery, has had that machinery produced by some component of human labor.

In the final analysis, there are only two components of any manufactured item. One is the human investment of time and energy, the other a completely raw material such as ore or rock. When a machine is used in the manufacturing of some parts, the machine encompasses the human labor of many hours. This labor, in one viewpoint, is stored in the machine and has its value added to each of the parts as it is manufactured.

# CHAPTER 16

# IS THIS A GOOD JOB FOR A ROBOT?

Some tasks are well suited for the use of robots, some are not. Table 15.2 in the previous chapter shows that, for the most part, the greater the number of parts and the more uniform the parts, the greater the advantage of automated production. There are many other factors to consider in the potential use of robots. The task of the robot application engineer is to use good judgment in deciding the merits of the particular installation for robot use. The following are some of the questions the application engineer might ask when investigating the job.

## ROBOT APPLICATION QUESTIONNAIRE

1. How many times will this operation be repeated?
2. How often are changes in the work anticipated?
3. What level of uniformity is there to the work?
4. What return on investment will be acceptable to the customer?
5. Can this job be done safely by a robot?
6. If the robot stops working, will the rest of the production stop?
7. Is there enough time to change over to the robot?
8. Is there enough floor space for the robot and its related equipment?
9. Is there overhead clearance for the robot?
10. Can the robot working area be enclosed?
11. Is there a proper power supply available?

12. Are trained application engineers available in the plant?
13. Are trained operators available?
14. Are trained programmers available?
15. Are trained maintenance workers available?
16. Is the process controllable by the robot?
17. Is the existing gauging system adaptable, if needed?
18. Is the robot able to reach all the points to do the job?
19. Is the surrounding equipment adaptable to the robot's features?
20. Is the time cycle long enough for the robot to handle?
21. Is a robot available with sufficient accuracy for the task?
22. Is a robot available to handle the weight of the work?
23. Is a robot available with axes able to produce all the desired motions?
24. Will the robot be able to handle the required accelerations?
25. Are there standard gripper components available?

Once questions like these are answered, the potential robot user must begin the task of selecting a robot with the best combination of abilities and price. Does the vendor have a reputation for quality and good service? Will the product be upgradable in the future? So many companies are producing robot equipment today, the selection process can be a long one.

To aid in understanding a good robot installation, let's examine a part produced by manual means and follow the design processes in its conversion to robot production.

## THE CONVERSION OF A TASK TO A ROBOT WORK CELL

The part we will manufacture is shown in Figure 16.1. It is a steel part that must be turned on a lathe, cut on a mill, located on a fixture, and welded into place. Used in production are two input conveyers, the welding gun, and an output conveyer. Figure 16.2 shows the general layout of the equipment.

### Specific Tasks

The person currently doing the job must do the following:

1. Pick up a part from the input conveyer.

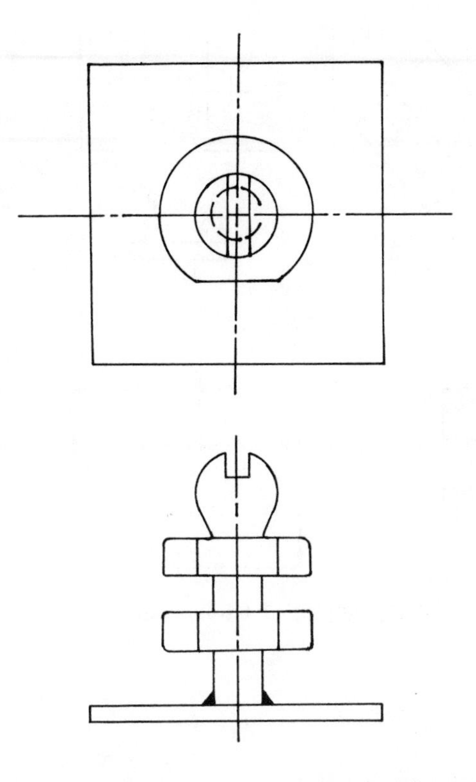

**Figure 16.1.** A part to be produced.

2. Wait for the lathe to stop and exchange the new part for the finished one.
3. Wait for the mill to stop and exchange the lathed part for the finished one.
4. Pick up a base plate from the conveyer and place it on the fixture.
5. Place the milled part on the base plate in the proper position.
6. Pick up the welding gun.
7. Weld the parts together.
8. Put down the welding gun.
9. Pick up the finished part and place it on the outgoing conveyer.
10. Start up again.

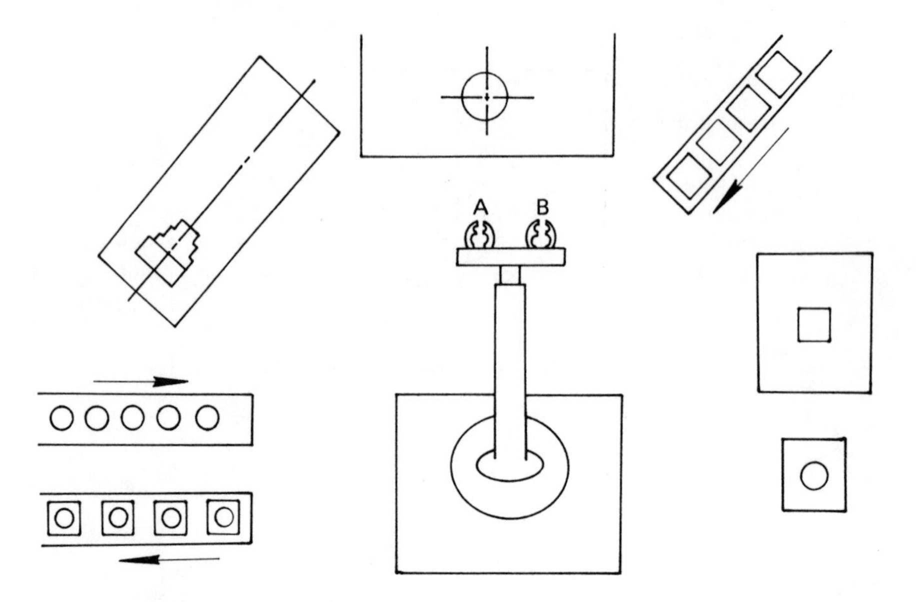


Figure 16.2. A robot work cell.

## Other Determinations

In addition to going through the basic motions, the human worker must make the following determinations:

1. Are there any parts on the conveyer?
2. Is the lathe really stopped?
3. Is the part stuck in the lathe?
4. Is the new part properly clamped in the lathe?
5. Is the mill really stopped?
6. Is the part stuck in the mill?
7. Is the new part properly clamped in the mill?
8. Are there any parts on the base plate conveyer?
9. Is the base on the fixture properly?
10. Is the part on the base properly?
11. Is the welding gun on the gripper properly?

12. Is the welding gun functioning properly?
13. Is the welding gun turned off?
14. Is there another part in the way on the output conveyer?

### Controlling Other Equipment

For a robot to take over the task of producing this part, we must also consider the way the human operator controls the other equipment.

1. When the lathe is loaded, push the start button.
2. When the lathe is stopped, release the chuck.
3. When the lathe has a new part, clamp the chuck.
4. When the mill is stopped, release the clamp.
5. When a new part is loaded, actuate the clamp.
6. When the mill is loaded, start the mill.
7. When the baseplate conveyer is empty, signal the foreman.
8. If the base plate does not fit the fixture, signal the foreman.
9. If the part is improperly located on the base plate, signal the foreman.
10. If the conveyer taking the parts away is full, signal the foreman.
11. If the input conveyer is empty, signal the foreman.

### Accomplishing the Tasks

This is the basic job we want to automate. Some of the actions are natural for the human worker, but they must be carefully delineated for the robot. The program the robot will use for each of these tasks will break the tasks into the same three groups outlined above. These are: actions, inputs, and outputs. Next, we will examine how each of these groups of tasks can be accomplished by the robot.

To pick up the parts and exchange them at each work station quickly, we can design a gripper with two "hands" very much like human hands. Figures 16.3 through 16.5 show the various shapes the gripper must fit in addition to the finished part. The motions that will take the gripper to the proper positions are shown in the sample section of the program that follows.

With one lathe part in side B of the gripper and the signal from the mill that it has completed its last operation:

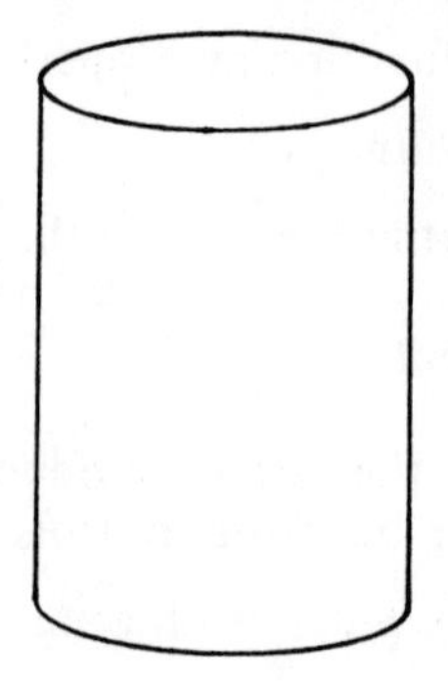

**Figure 16.3.** Original shape supplied to robot.

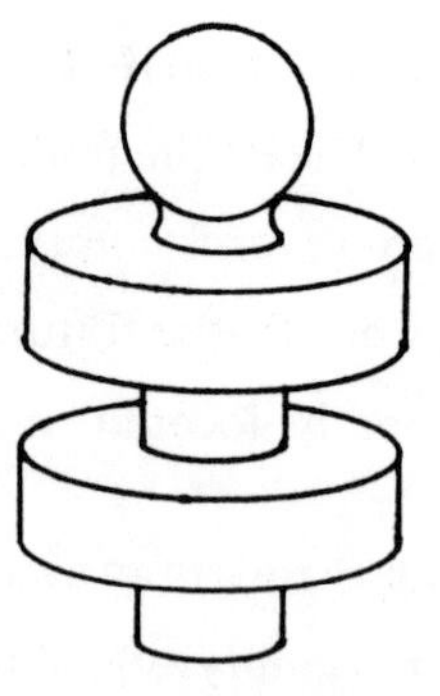

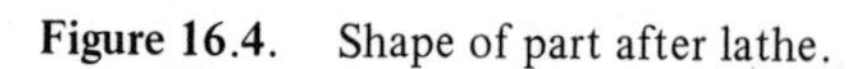

**Figure 16.4.** Shape of part after lathe.

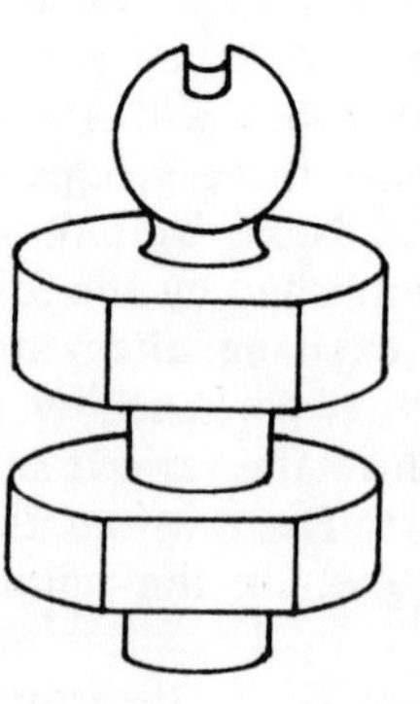

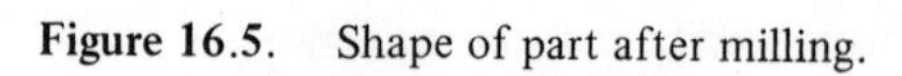

**Figure 16.5.** Shape of part after milling.

step 300 Go into mill
step 301 Line up with chuck
step 302 Command gripper A to open
step 303 Place gripper A over part A
step 304 Close gripper
step 305 Signal mill to open chuck
step 306 Check to see if chuck is open
step 307 Remove part
step 308 Check gripper for the presence of the part
step 309 Rotate gripper
step 310 Place part B in chuck
step 311 Signal chuck to close
step 312 Check to see if chuck has closed
step 313 Release gripper
step 314 Move gripper away
step 315 Check gripper to see if part is present
step 316 Move to next work station

When the application has a blocked out program like this one (usually every line would not be written), it can be determined if the robot will be able to accomplish the same physical labor as the human worker. What may still be unclear is the robot's ability to communicate with the other equipment.

## Communication

Lines 306, 308, 312, and 315 are examples of inputs from the other equipment to the robot. Lines 302, 304, 305, 311 and 313 are examples of outputs from the robot to the other equipment (the gripper itself is sometimes considered outside equipment). Figure 16.6 shows the type of connection on the other equipment that will allow the robot to communicate. Once the equipment has been checked for the adaptability of robot communication, the application engineer starts to get a good idea if the job will be easily accomplished by a robot. Factors which must now be considered are: the robot's speed, safety requirements, and problems of maintenance.

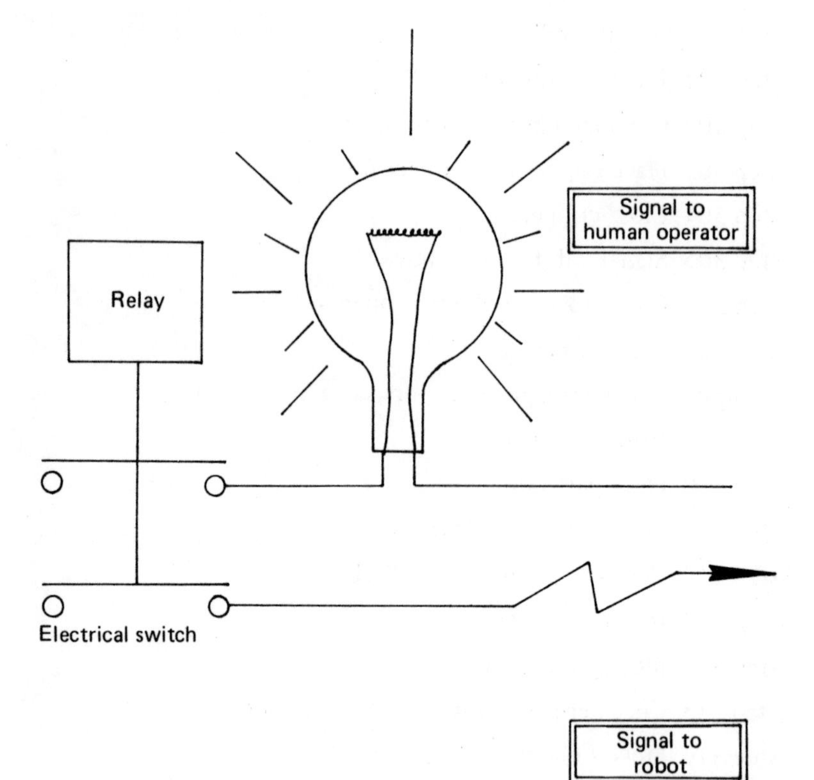


**Figure 16.6.** The type of signal sent to the robot.

## Speed

If the job to be done is one of rather slow pace, there is little concern about the robot's speed. If the job is to be done very slowly, however, we would most likely not be bothering to use a robot. Unless the actual process can be simulated (perhaps by computer or by a robot in the laboratory), the time study of a robot installation can be a very frustrating part of the application engineer's job.

Table 16.1 shows the machines and processes involved in the work cell. The calculations do not predict how long the robot would take to do its work, but whether the robot could do the work in the time allowed for the other processes. If the robot were faster than the other machines, then the overall time cycle would be proceeding

**Table 16.1. Robot Time Cycle Calculation**

| Travel/Function | Distance | Normal Robot Speed | Time (sec.) |
|---|---|---|---|
| Output to input | 34 in. | 45 in./sec. | 0.75 |
| Input to lathe | 61 in. | 45 in./sec. | 1.35 |
| Lathe to chuck | 11 in. | 40 in./sec. | 0.275 |
| Grip | | 0.4 sec. | 0.40 |
| Chuck response | | 0.5 sec. | 0.50 |
| Rotate gripper | 39 in. | 40 in./sec. | 0.975 |
| To chuck | 4 in. | 45 in./sec. | 0.09 |
| Chuck response | | 0.5 sec. | 0.50 |
| Release | | 0.4 sec. | 0.40 |
| Out of lathe | 11 in. | 40 in./sec. | 0.275 |
| Over to mill | 68 in. | 45 in./sec. | 1.51 |
| Etc. | | Total this section | 7.03 |
| | | Rule of thumb for starting and stopping motions + 100% | 7.03 |
| | | Allow for this section | 14.06 |

at maximum with that equipment. The robot might still be used, even if it were the slowest of the machines, as long as it could keep enough parts produced to satisfy some standard. (Perhaps the next series of machines could only produce at a given rate, and the robot would have to wait if it were any faster.)

### Safety

Could this job be made safe with the use of a robot? Figure 16.7 shows the basic machine layout and a barrier that would be installed to protect the human operators in the area. If for some reason, a person would go into the enclosure, the robots controls would be so wired that the robot would stop as soon as the door was open. With these two safeguards, the installation could be said to be safe.

### Maintenance

As this robot would be used, the parts would be fed to it on a conveyer and would be taken away on a conveyer. The machines in the next work group would perform additional work on this same

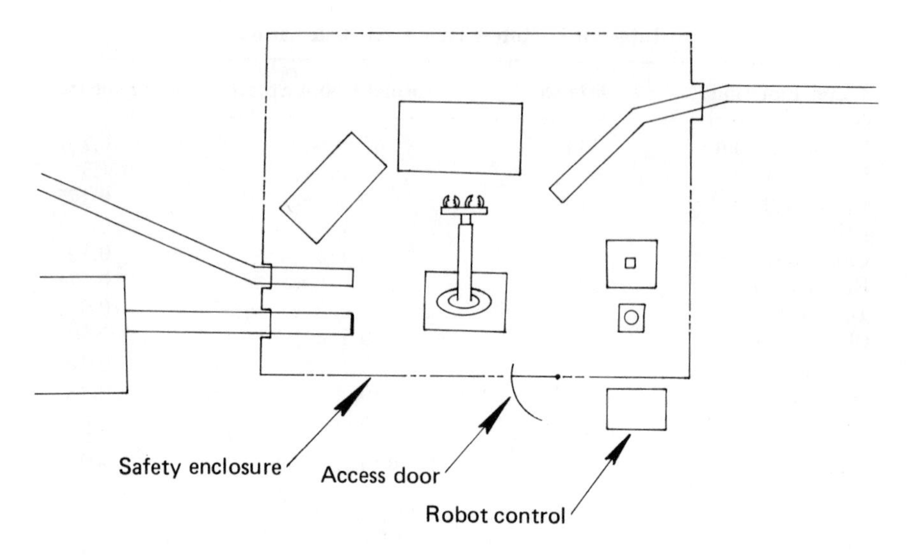


**Figure 16.7.** Safety enclosure for the work cell.

part. If the robot malfunctioned and had to be stopped for repairs, what would happen to the work of the other machines? If the conveyer were short, there may be only a few parts on it at a time. When repairs would take more than the time these parts would take to be processed, the other machines would be forced to stop due to lack of parts.

In a continuous-path assembly line, the breakdown of even a single work cell can cause a disruption in the entire operation. While the value of the single cell in terms of output-per-hour may not be so great as to cause alarm, the output of the entire plant may have a value so great that even the loss of a few hours' production is more than the cost of a new robot.

As one plant worker put it, "In the old days, we had men working shoulder to shoulder on the assembly line. If a man had to stop because he was sick or something, the relief man just came over and filled in on his job."

Robots need to be *repaired* when they are "sick." It is not economically feasible to have a robot on hand to take over the work of a broken one. Our installation should have little problem with maintenance, however, because there are several hours of parts between each of the work stations in the plant.

### Profitability

We know that the robot can actually do the work we have in mind and that the work can be performed safely and with few maintenance problems. But, would this installation be a profitable investment? The formulas for determining the cost factors of an installation were presented in the previous chapter. Let's just say that a work cell of this type might run two parts from the robot for every one part from the human operator. If there is sufficient market to justify a second and third shift, then this installation would have a very short ROI.

## CHAPTER 17

## SOME HOPES FOR THE FUTURE

In the previous chapters, we have been discussing practical industrial robots currently in use. In the future, the robot industry is expected to grow faster than any other. There are jobs and processes now only dreamed of that will be commonplace for the robots of the future. The following are some examples of things robots may some day do. This short list is just the tip of the iceberg for robot potential.

### A FIRE-FIGHTING ROBOT

Imagine this scene: Responding to a call for help, firemen arrive at a burning building. Flames are present at all the windows on the first floor. The building will soon be engulfed in flames. The risk to the safety of the fire fighters themselves does not allow them to even attempt to rescue someone trapped on an upper floor.

Fortunately, on the back of the fire truck is a robot fire fighter (Figure 17.1). Responding to a signal from the firemen, it leaps off the truck and approaches the house. The robot smashes through the door, enters the house and, following preprogrammed instructions, searches the house for victims. Special metal eye strips that are applied to door frames signal the robot of rooms occupied during the night or occupied by those not able, by themselves, to escape from a fire.

Upon entering one of the inner rooms, the robot sees a man overcome by smoke, yet still alive. There are only moments before the building will collapse in flames. The robot grasps the victim, opens the front cavity of its robot body, and places the victim inside. The chamber is air conditioned and has its own oxygen supply. Once the door is closed, it insulates the victim from outside temperature extremes. A small TV screen allows the victim to communicate with the

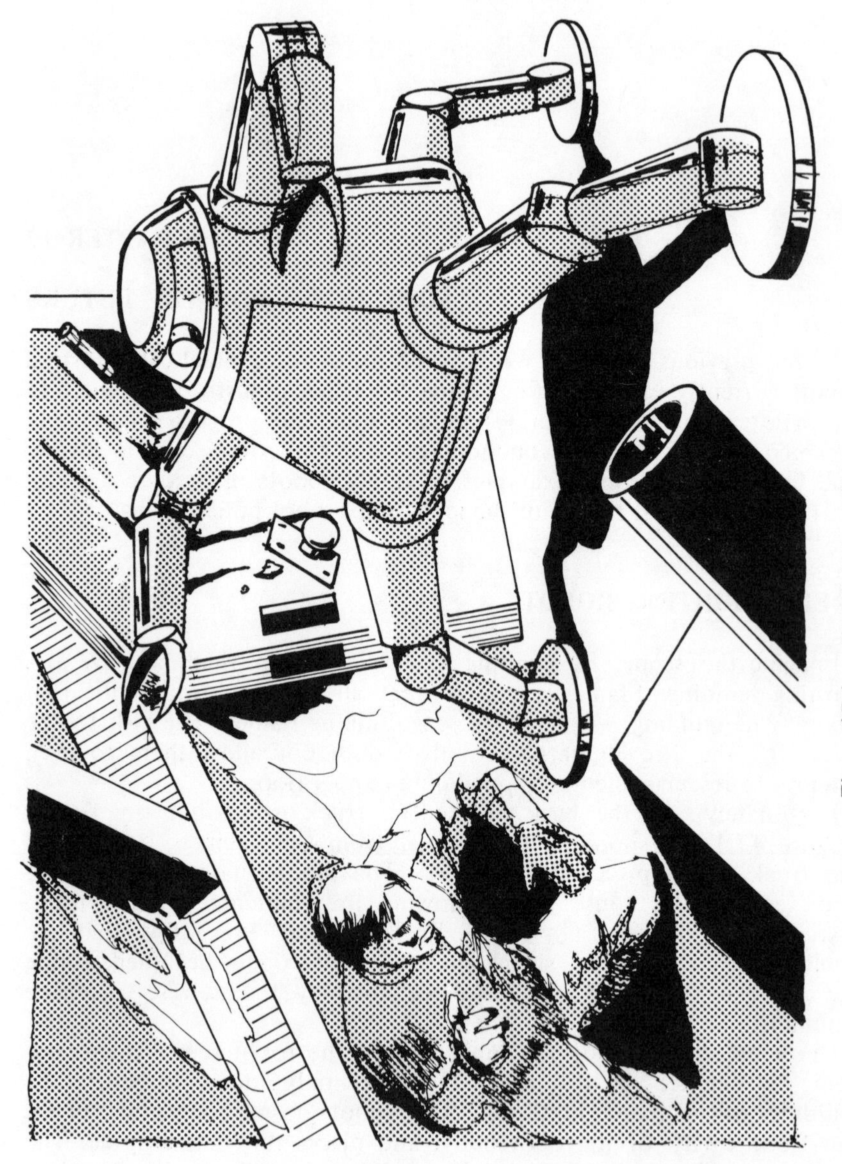

**Figure 17.1.** Fire-fighting robot.

firemen outside. The victim's pulse, blood pressure and breathing are all monitored; and, when the robot takes him to a place of safety, paramedics are already present to provide for his care.

The robot is equipped with legs to carry it about; but, once it has arrived on the scene and has proceeded to protect a victim, a malfunction in the legs or the collapse of the building would not jeopardize the safety of the rescued person. The robot could simply wait some minutes while rescue workers sent another robot to the aid of the first or used a crane-type of machinery to lift the now immobile robot from the burning wreckage.

Because the robot is only a machine and—although costly—a replaceable resource, this robot fire fighter could be used in ways that a human fire fighter never would. If a building is on fire and there are no human beings within, a robot fire fighter might still enter the building to put out the fire or to rescue valuable objects. Robots used by a particular local fire department could be preprogrammed with instructions for each individual dwelling and business place in the area. If an office building housed the studio of an artist whose one prized possession was a priceless old master, the robot could, when all humans had been saved, retrieve this painting.

A fire fighting robot, because it would be impervious to open flame and high temperature, could carry chemical fire extinguishers to the actual point of ignition. The practice of pouring water on one part of a building to extinguish a fire unreachable by a direct stream of water would no longer be necessary. This would be of great benefit because some of the greatest fire damage is caused, not by the flames themselves, but by the water used to extinguish the flames.

## SHIP VAC: A SHIP-HULL-CLEANING ROBOT

Large ships must occasionally go to a special dry dock and be removed from the water to have their hulls cleaned and inspected. Figure 17.2 shows a robot with special magnetic feet walking along the underside of a ship while it is in the water. This robot has the ability, with a vacuum nozzle, to scrape and clean this underside. Special testing equipment with a highly accurate sonar device probes the metal thickness as it cleans. If any particular section of the ship has unusual wear, the robot so alerts a repair crew.

Because the robot can work without any human attention, it can function at night in a crowded harbor or while the ship is underway without fear of danger to personnel. The equipment can also be used to perform minor repairs such as the welding of leaks and the fitting of propellers. The ship vac is of little interest to a shark or other predator

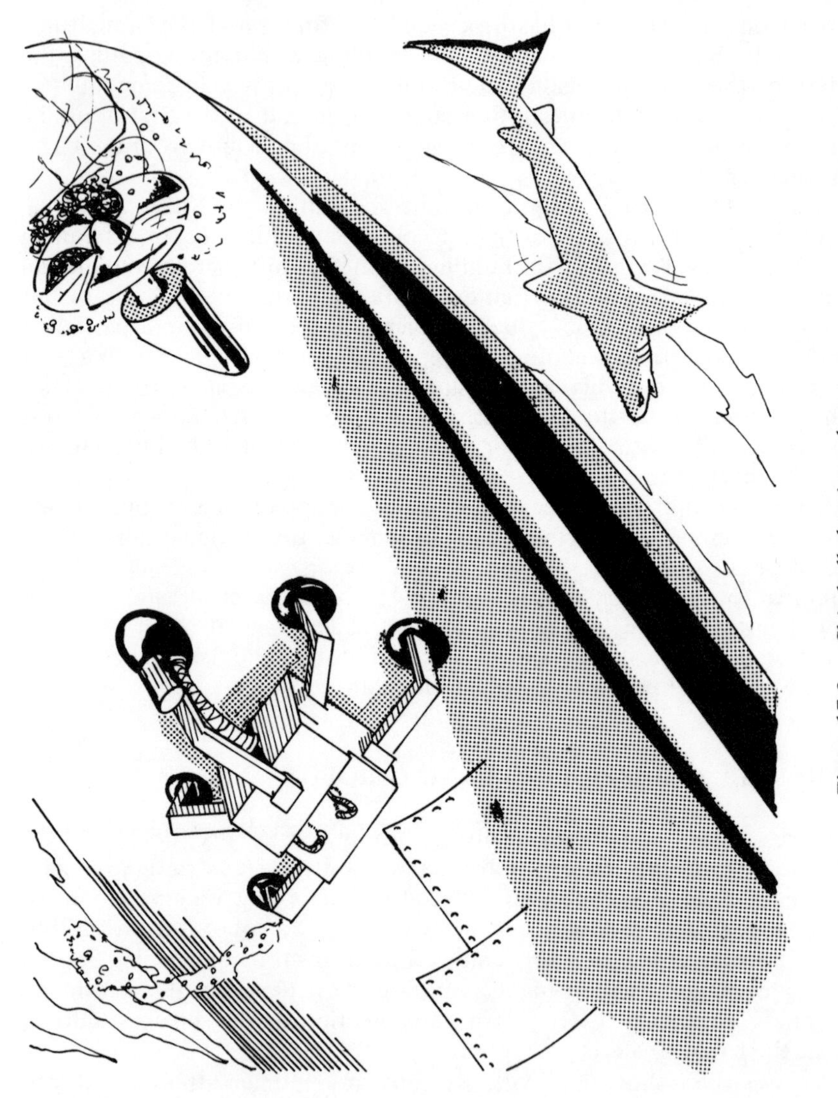

Figure 17.2. Ship-hull-cleaning robot.

of the deep. What's more, the robot can carry out its task under any condition of dirty water.

## ROBOTS FOR NUCLEAR MAINTENANCE

Deep within the reactor core of a nuclear power plant, we find special radiation-hardened robot workers (Figure 17.3). When the reactor was built, the robots were placed within before the reactor was sealed. They are called on routinely to service the reactor components and to make repairs. It is not necessary to power-down the reactor and go through an extensive decontamination process each time there is a minor malfunction or potential danger. Specially designed robots which can withstand very high radiation exposure do the work within the reactor core while the reactor is in process. Potential dangers can be quickly corrected by the robot.

To avoid the expense of many millions of dollars to shut down and decontaminate a reactor, several robots and a large amount of robot spare parts are provided as standard equipment in the reactor. Each robot is designed and built so that it may remain idle for a long period, and yet immediately be brought into use. If one robot malfunctions, there is another to take its place. A second robot can even perform, under the watchful gaze of its human directors, repairs on the first robot.

A special covering is provided with the robots so that when the reactor chamber is eventually opened, this cover can be removed and discarded along with other radiation-exposed materials. Outside of the reactor chamber itself, where nuclear fuel is processed or stored, robots can relieve human beings of the potentially lethal task of handling the radioactive materials.

## BRIDGE-PAINTING ROBOTS

On some large-span bridges, there are painting crews working continuously. The crews work hard and fast. But, because the surface area of a bridge is so great, by the time it is completely painted, the portion painted first needs to be painted again.

Figure 17.4 shows a pair of bridge painting robots going about their work. Able to hang at any angle and to carry themselves around the bridge without the aid of scaffolding or safety lines, the robots can paint a large section in a short time. The tanks within the robots carry sufficient paint so that they may work for hours without returning to a tending vehicle.

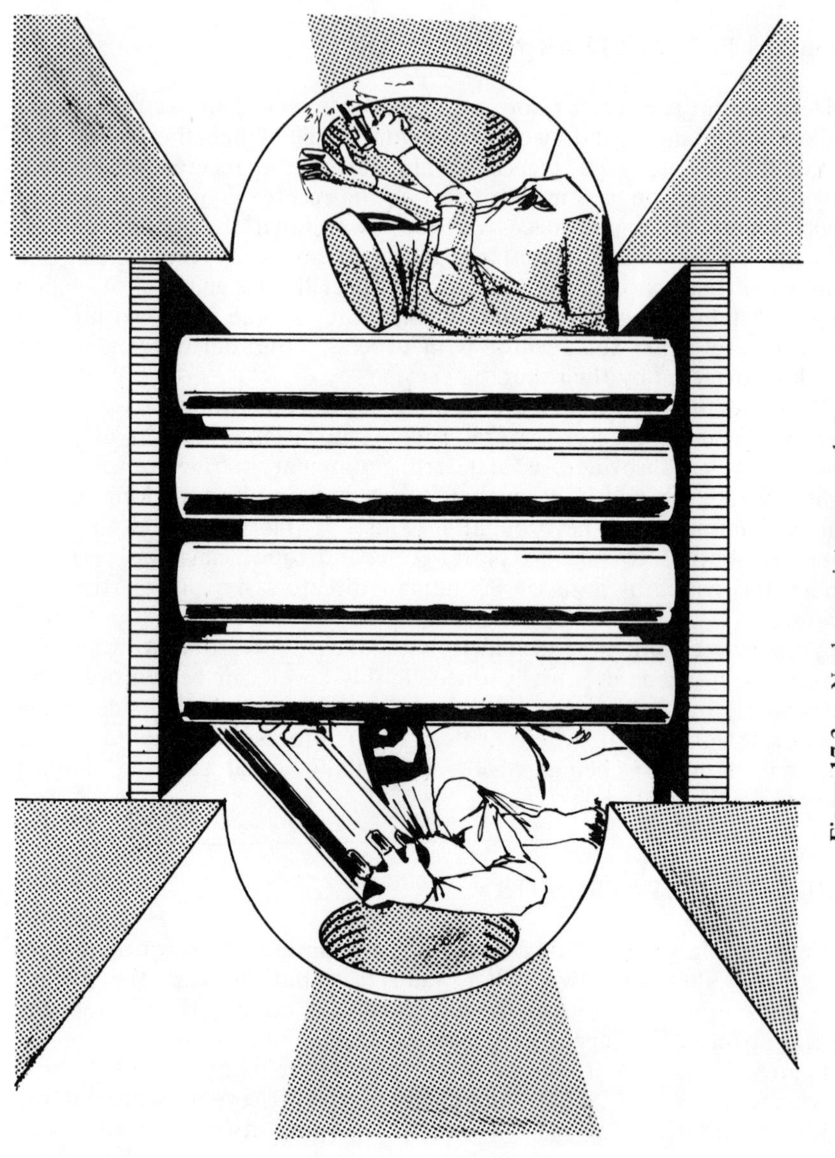

**Figure 17.3.** Nuclear maintenance robot.

Figure 17.4. Bridge-painting robots.

If some malfunction occurs, a robot can, at a time of safety, release its grasp and fall harmlessly into the water, where it can be retrieved and put into the contractor's boat. If necessary, a spray-painting robot can even swim out to an intermediary support and climb into position itself. Each individual bridge does not need its own set of spray-painting robots. Because the work is accomplished quickly, robots move from bridge to bridge, reducing the overall cost of bridge maintenance.

## THE ROBOT SEWER WORKER

Popular fiction about grotesque animals aside, sewers are dirty places. There is legitimate danger in the myriad tunnels used to carry waste. Over long periods of time, the exposure of organic material to conditions of unnatural growth can create an atmosphere both full of quasi-toxic substances and depleted of oxygen. Aside from the obvious source of biological contamination, the sewers of a large city contain many things hazardous and unpleasant.

Figure 17.5 shows a specialized bulldozer type of robot which can travel in the sewer, find its way automatically through the maze of connections and corridors, and dislodge obstructions. During times of flood, the robot can act as a ram to push water like an oar, facilitating extra water removal by the system. A specialized underground garage is the entry point for such a device, allowing it to travel under the streets and buildings of a city while carrying out its work.

## SMOKY-THE-BEAR ROBOT

One reason that Smoky the Bear is used as an example of American forest fire fighting is his proximity to the forest. Unlike a bear, the ordinary forestry fire fighter is not close at hand when a fire starts. It is, of course, very dangerous for a human to be close to a forest fire, especially when there is a limited possibility of leaving the area.

As with our home fire fighting robot, the Smoky-the-Bear robot (Figure 17.6), while working automatically, is impervious to the dangers it encounters. Although the robot itself might be damaged by a falling tree, flames and high temperatures themselves do not cause it harm. The large tank-like robot carries a large quantity of water and automatically distends a suction hose, allowing it to draw water from a lake or stream. Its high pressure pump shoots the water exactly at the point of flame and eliminates the need for haphazard bombing by airborne fire fighting equipment.

The helicopter in Figure 17.6 is another robotic device. Since it is flying in an unpopulated area and is exposed to some danger, the

Figure 17.5. Robot sewer worker.

**Figure 17.6.** Smoky-the-Bear robot.

helicopter is an ideal candidate for a robotized function. Special programming is created in advance of a fire indicating by type of tree, geographic location, wind velocity, fire origin and spread the exact positions where water cannons should be placed. Information is relayed by the ground-based and air-based equipment. In places where no man would go, the robot workers can gather valuable information.

## A ROBOTIC ARTIFICIAL HAND

Although we have said earlier that human features for robots are an element of science fiction, Figure 17.7 is an example of an almost lifelike robotic hand. The limb itself is coated with a plastic and lifelike skin, and inside is a series of microrobotic components. A signal from the wearer causes automatic functions in the robot hand. Motions as simple as grasping a coffee cup or as complex as playing the piano are programmed into the robotic hand. Although there is no substitute for real flesh and blood response, the disabled person, when fitted with a prothetic device manipulated by robotic components, should gain much of the ability originally lost.

## A MINE-WORKING ROBOT

Since man's discovery that ore could be turned into metal, he has dug holes in the ground and brought ore to the surface. The recovery of material from deep within the earth is sometimes at great cost. Even though the modern mine is, for the most part, a safe working environment, there are accidents and injuries. With special robotic equipment (Figure 17.8), the miner remains on the surface in a supervisory capacity and allows the robotic equipment to descend into the mine where hazardous conditions such as explosive gas, toxic gas, or frail rock formations are present. The robotic miner proceeds just as though conditions were perfect.

Ore and minerals are retrieved that formerly were economically unfeasible to mine. Even when a human miner must work within the mine, the robotic mining equipment aids in the digging process. The old song about sixteen tons gives rise to a new level of mining performance. Just as in the modern industrial plant, the proper use of automation equipment drastically improves the profitability of the job.

## A CONTAMINATION-PROOF NURSE

In extreme cases where a hospital patient must be kept isolated and free from contamination from the medical staff, a robotic nurse

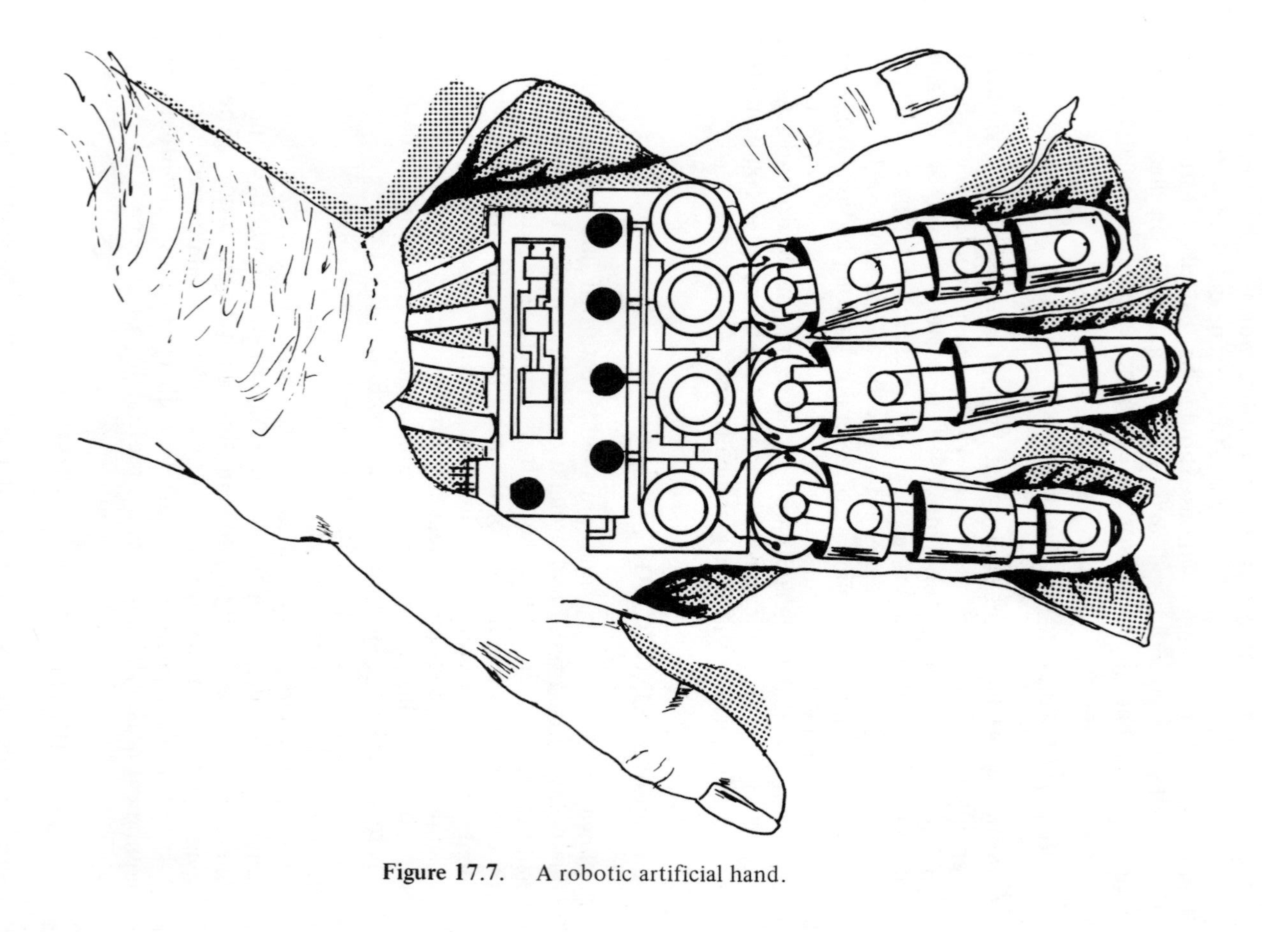

**Figure 17.7.** A robotic artificial hand.

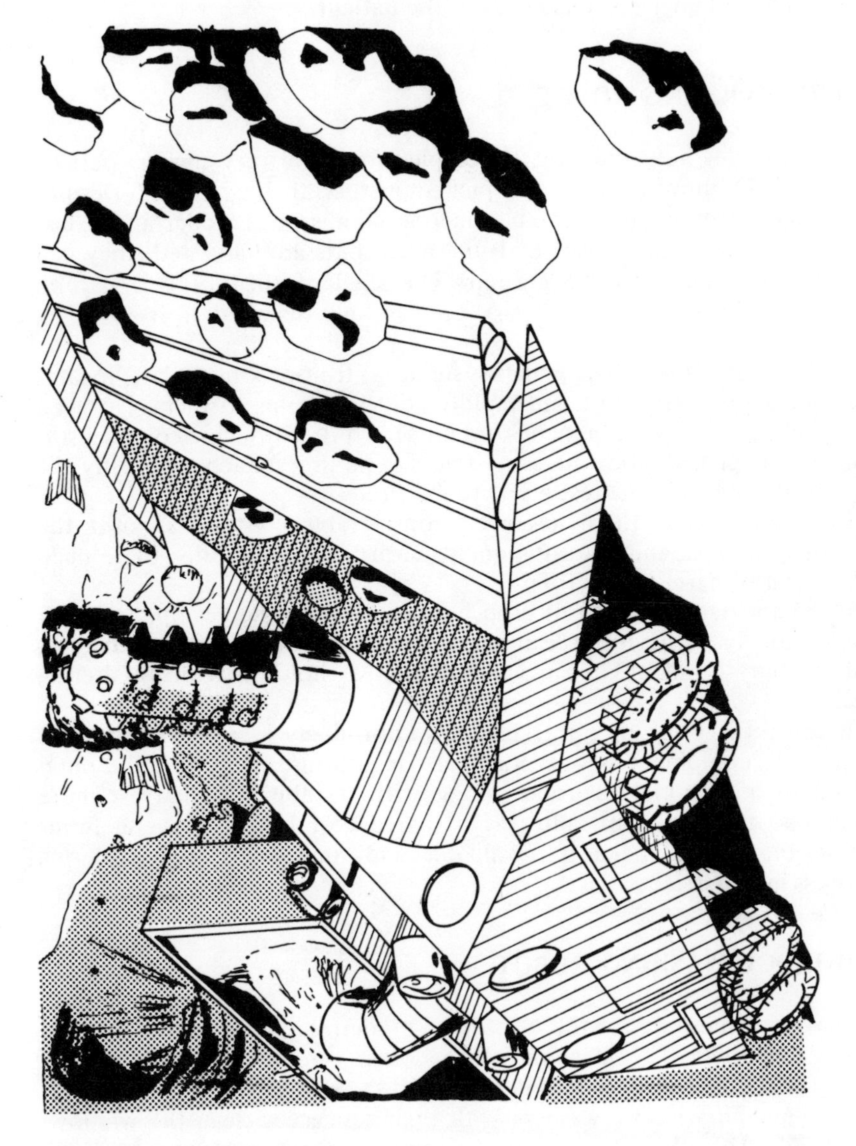

Figure 17.8. Mine-working robot.

(Figure 17.9) can provide patient care. Although there is no substitute for the personal contact between doctor and patient, no substitute for the healing power of a nurse's touch and warmth, a robot nurse fulfills the simple and unglamorous needs of the patient.

## A FRUIT-PICKING ROBOT

Ever wonder how they get the apples from the very highest part of the tree? Or how cherries are picked just as they get ripe? Despite mechanical fruit pickers, often someone climbs on a ladder and painstakingly pulls at each piece of fruit. When nuts are harvested, they can be allowed to fall to the ground. The shell of the nut is a natural protector. Fruit, however, must be carefully handled if it is to be saleable.

With a rapid-vision-analyzing system, a fruit-picking robot locates individual pieces of fruit and rapidly guides its arms to them. No need for ladders, no broken branches, no missed fruit. Through a color sensor, the fruit is picked from the same tree several times. Each time, only the ripe fruit is picked–the rest is left to mature.

In recent times, there has been considerable controversy over the use of pesticides and the subsequent harm to workers. But, there need be no fear of harm to this robot.

It is important that a fruit crop be speedily harvested as ripe fruit can spoil quickly. This robot can work 24 hours-per-day, it can harvest throughout the night, it can work in rain or hot sun with unfaltering ability.

Because the price of farm labor is so crushingly low, most expensive automatic equipment is not practical for farm use. However, the rapid speed of a fruit-picking robot along with its ability to harvest more than one type of fruit–allowing a shared cost between several farms or regions–could make it a valuable addition to the food-gathering process in America.

## A WINDOW-WASHING ROBOT

Glass is often used on the outside of modern high-rise apartments and office buildings. The cleaning of this glass can be a demanding and sometimes dangerous job. Perched on a scaffold high above the street, human window washers with their squeegees clean the windows by hand. An automatic window washing robot can clean and repair glass quickly and efficiently. Alas, stories of window washers and what they see in windows would end if window washers were machines instead of persons.

Figure 17.9. Robot nurse.

## A ROBOT MAINTENANCE WORKER

Question: How many robots does it take to change a light bulb?
Answer: One, but it costs $1,000,000. What sounds like the beginning of an ethnic joke is in reality a statement about the potential ability of robots.

A robot to change the light bulb in a secretary's desk lamp is unreasonable. However, our fantasy robot reaches far into the air and quickly exchanges light bulbs in an industrial plant, saving considerable money. The robot is also valuable because the light bulbs are of mercury vapor or other such toxic chemical, and bulb damage could expose a human worker to some harm. The automatic light-bulb-changing robot maneuvers on freeway overpasses and in underground tunnels, eliminating the necessity of humans working in these hazardous sites.

## ROBOT GOLF CADDY

The occasional golfer who finds himself on a course with a lost ball and little help of salvaging a normal score for the day should appreciate the robot golf caddy.

Through a small radio transmitter on the player's belt, the caddy automatically tracks and follows the player throughout the course. When the ball is struck, the special tracking apparatus within the robot follows the ball and notes the exact position to the player. Should the player request advice about the angle, distance, or ground conditions at a particular point, the caddy scans its memory of course topography and answers questions. Wind velocity, temperature and humidity conditions are considered in the advice.

If a player feels this information would take all the fun out of the game, the robot can be instructed not to give any advice. There is, of course, the possibility of the caddy striking the ball to get the player out of a jam. But here, we will draw the line. Robots have their place in many of our daily activities, but when they begin to get better golf scores than their human masters—particularly if humans start to carry the clubs—then they have gone too far.

## STILL FANTASIES

Although none of these robot fantasies—from the fire fighter to the golf caddy—is yet possible with today's technology, they may in time become realities.

## CHAPTER 18

# A SIMPLE ROBOT PROJECT

Notes and figures on a simple robot project are included to illustrate the simplicity of some robots and encourage experimentation. The information is not intended to be a "cookbook," build-it-yourself set of directions. Dimensions and construction techniques are specifically avoided. Rather, what follows is a list of simple components and drawings of a robot put together by the author (Table 18.1 and Figures 18.1 through 18.8).

The robot is used to lift a small object. It lifts the object, rotates, extends its arm, and sets down the object. The control mechanism is built around a valve actuator that is commercially available. When this robot is run, the cycle repeats itself continuously until the air supply or the electric supply to the valve actuator is turned off.

The robot is composed of six air cylinders and their associated mechanical linkages. The base unit has two pipes that serve as a bearing. As the first cylinder extends, it causes the upper unit to rotate.

The two middle cylinders lift the upper portion. Two cylinders are used to restrict further rotation of the upper unit. The top portion has two cylinders that extend the gripper. The gripper itself needs only one cylinder to grasp any small object within the working limits of the robot.

The control mechanism is a simple connection of valves to the cylinder. The cam-operated valve controller can be programmed to make any combination of motions desired.

Try developing your own robot. If the future follows the pattern of the past, someone who is just now learning about robots may become the next Edison or Ford.

**Table 18.1. Components of a Simple Robot**

| Item Number (keyed to figures) | Description | Author's Cost |
|---|---|---|
| 1 | Plywood rotational platform | $ 2.00 |
| 2 | Outer pipe bracket | 2.59 |
| 3 | Rotation cylinder rod clevis | 2.50 |
| 4 | Outer pipe | 1.85 |
| 5 | Inner pipe | 1.65 |
| 6 | Bolt | 0.59 |
| 7 | Inner pipe bracket | 2.39 |
| 8 | Mounting surface | 0.00 |
| 9 | Lift cylinders (2 @ $35.00) | 70.00 |
| 10 | Plywood lift base | 2.00 |
| 11 | Nut (4 @ $.22) | 0.88 |
| 12 | Reach cylinder (2 @ $25.00) | 50.00 |
| 13 | Grip cylinder | 15.00 |
| 14 | Channel aluminum grip body | 4.25 |
| 15 | Grip body nut | 0.24 |
| 16 | Cylinder rest pad | 0.00 |
| 17 | Rod pivot pin | 1.00 |
| 18 | Rotation cylinder | 20.00 |
| 19 | Rotation support bracket | 2.00 |
| 20 | Rear pivot pin | 1.00 |
| 21 | Rotation cylinder rear clevis | 3.25 |
| 22 | Valve actuator | 35.00 |
| 23 | Actuator disks (4 @ $2.00) | 8.00 |
| 24 | Valves (4 @ $10.00) | 40.00 |
| 25 | Valve actuator motor | 25.00 |
| 26 | Assorted tubing and fittings | 15.00 |
|  | Total | $306.19 |

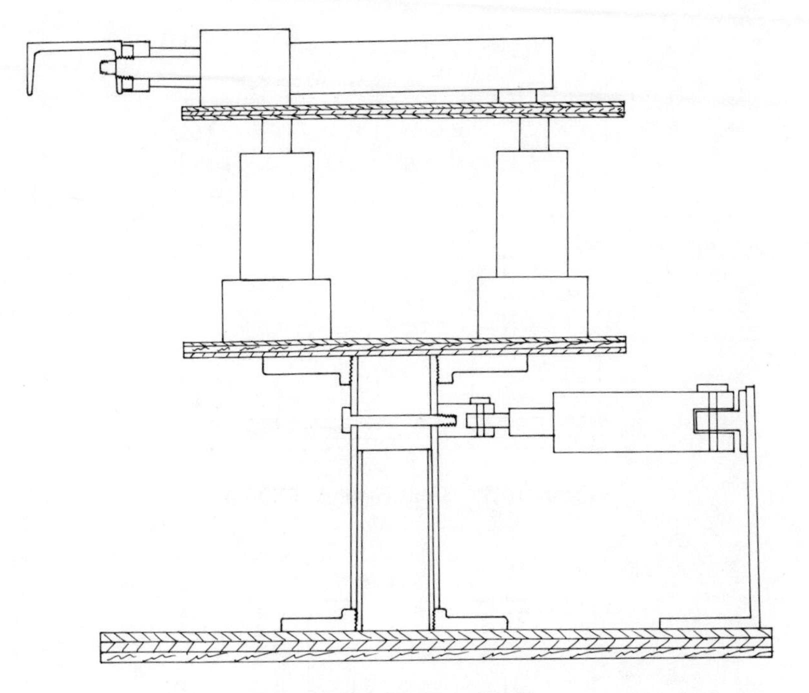

Figure 18.1. Robot assembly.

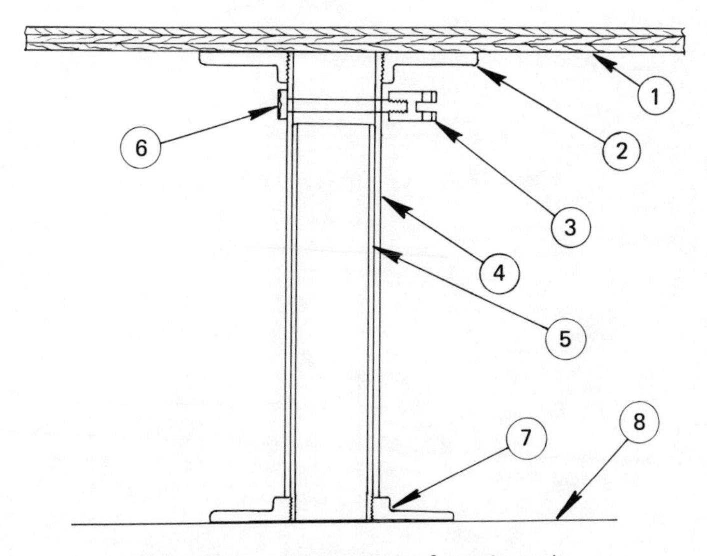


Figure 18.2. Construction of rotating unit.

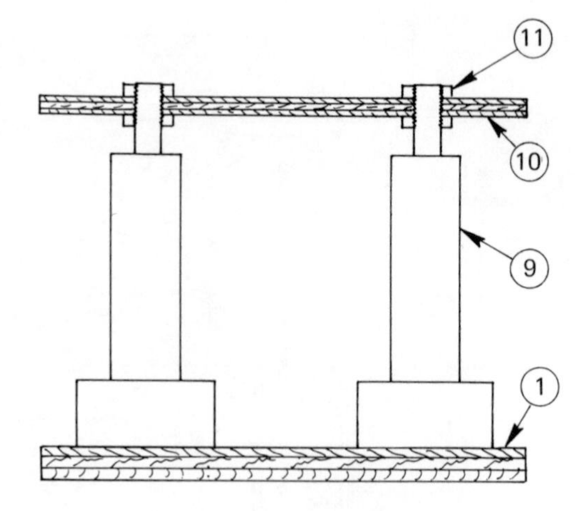


Figure 18.3. Construction of lift unit.

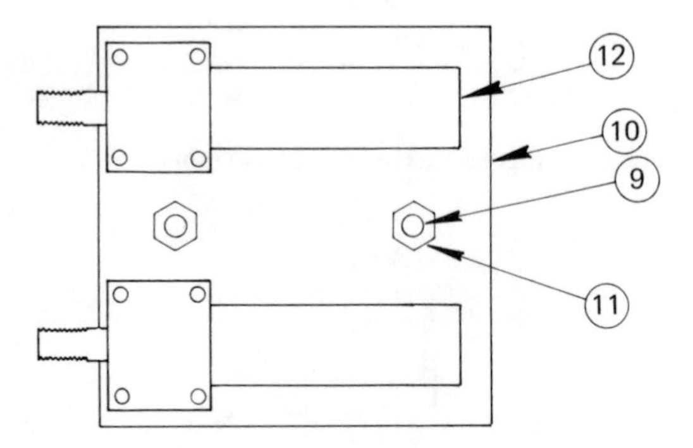


Figure 18.4. Construction of reach unit.

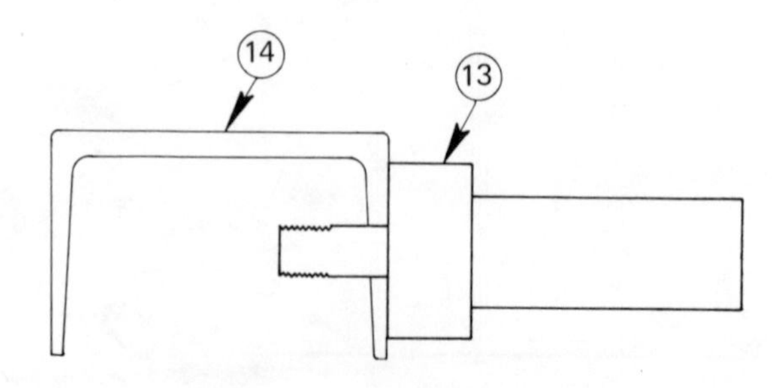


Figure 18.5. Construction of grip unit.

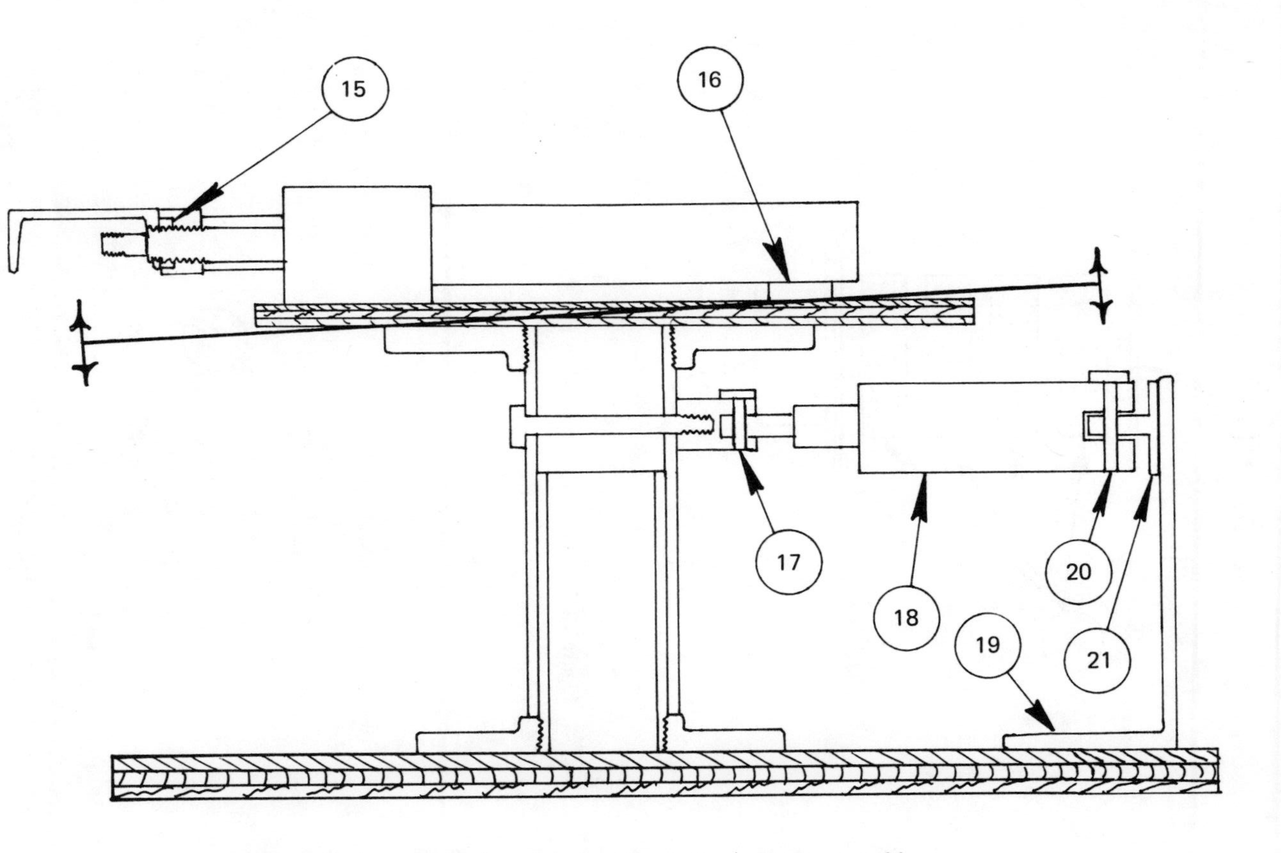


**Figure 18.6.** Detail of reach/grip and rotation/cylinder assembly.

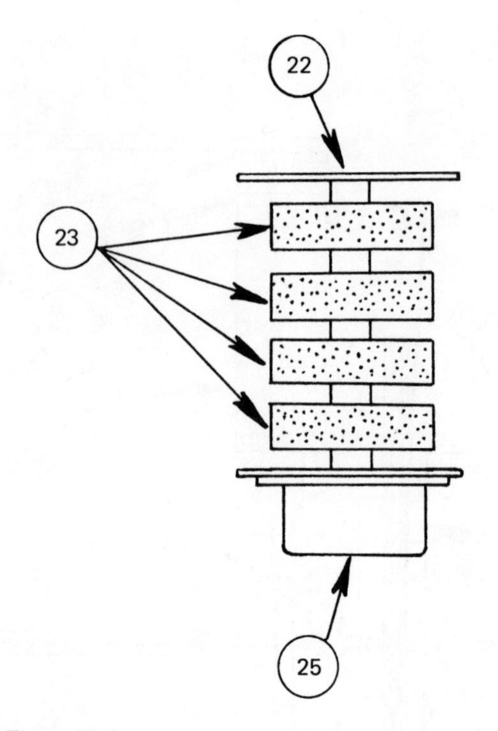


**Figure 18.7.** Valve actuator (valves not shown).

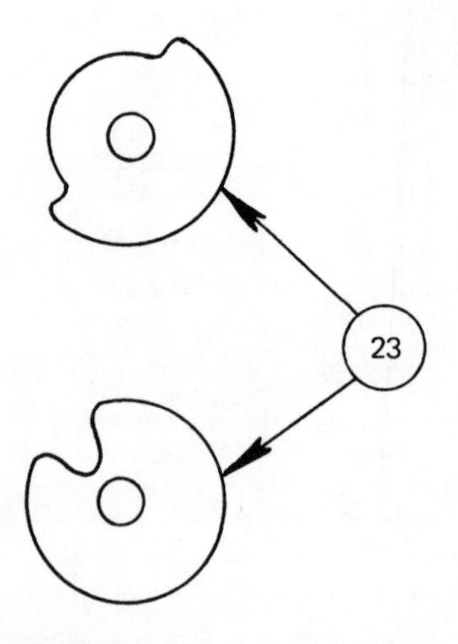


**Figure 18.8.** Detail of valve actuator cams.

# GLOSSARY

ACCURACY. The ability of a robot to reach a given point in space repeatedly, and how far off it will be in the worst case.

AIR CYLINDER. A piston in a cylindrical housing moved by pressurized air.

AIR LOGIC. Computer-type logic calculations accomplished by small air valves.

AIR MOTOR. A device with a shaft that will turn when pressurized air is applied.

ALGORITHM. A set of procedures used to solve a problem.

ANALOG. In electronics, a signal that gives information by a change in either its current or its voltage.

ANTHROPOMORPHIC. Manlike.

ARC WELDING. The process of joining metal together by applying electric current through a protruding wire, melting the pieces and the wire together.

AXIS. In physics, the center of a rotating member. In robotics, an individual component free to move relative to the other components.

BALL SCREW. A device for transforming rotary motion to linear, or vice versa, incorporating a threaded rod portion and a nut consisting of a cage holding many ball bearings.

BIT. One single piece of computer information, either *on* or *off*, 0 or 1.

BYTE. A single computer word. Depending on the hardware used, a byte may be 8, 16, 32, etc., bits.

CABLE CYLINDER. A piston within a cylindrical housing that has a flexible cable joined to both ends. As the piston moves, it causes the cable to move over wheels at each end of the cylinder.

CAM. A rotating part which, due to its eccentric center line, causes an in-and-out motion in any part pushing against it.

CAM FOLLOWER. A special lever with a wheel at its end designed to push against a cam with little friction.

CAPACITOR. An electric device used to store electricity temporarily.

CARD (COMPUTER). A printed circuit board.

CARTESIAN COORDINATES. A means for giving the location of a point in space by measuring its distance from a reference point along straight lines. The lines are at right angles to each other and all meet at the reference point. Named after its creator, René Descartes.

CHAIN DRIVE. Any device that transmits power by use of a chain and sprocket.

CHIP. A single electronic part, usually to be placed on a circuit board.

CHUCK. A set of clamping jaws.

COLLET. A series of projections held in a collar that may be tightened around a separate shaft.

COMPUTER BOARD. A printed circuit board used in a computer.

CONTACT DRUM. A cylinder with electrical conducting strips around its exterior that, when rotated, will cause the connection of electrical contacts.

dBA. A rating system used to measure the harmful effect of sound. dB equals decibels, A the specific scale (there are also B and C scales).

DEBURRING. The process of removing burrs from a part, also rounding sharp edges.

DEVIATION. The amount of variance.

DIAGNOSTIC. Information indicating the nature or location of a malfunction.

DIGITAL. In electronics, a signal that is either *on* or *off*; see ANALOG.

$E^2$PROM. Electronically Eraseable Programmable Read Only Memory.

EMULSION. A mixture of oil and water that does not readily separate.

ENCODER. An electronic device used to measure small units of motion, often using a light source and detector.

END-OF-ARM TOOLING. Virtually anything that is placed on the end of a robot arm.

EPOXY. A glue made up of two chemical parts.

EPROM. Electronically Programmed Read Only Memory.

EXECUTIVE MEMORY. The portion of memory holding the executive program.

EXECUTIVE PROGRAM. The basic operating instructions for a robot placed in memory at the factory.

FEEDBACK. A signal from the robot equipment about conditions as they really exist, rather than as the computer has directed them to exist.

FEEDBACK LOOP. A signal originating in the computer that is changed by the robot and given back as feedback.

FIBER OPTICS. The use of thin strands of flexible glass to transmit light around corners.

FLIP-FLOP. A logic element that will give one of two signals each time it is stimulated, alternating between the two.

FLOPPY DISK. A small flexible disk used to record computer information.

FLUORESCING. To give off very bright light when shone on by light.

FLOW-CONTROL VALVES. Valves that can change the rate of flow to equipment.

GEAR MESHING. The placing of two gears so that their teeth will fit together without interference.

GRIPPER. End-of-arm tooling used to grasp objects.

HALL EFFECT SWITCH. A switch that conducts when a magnet is placed close to it.

HARD-WIRED. A connection made directly with wires rather than through a computer.

HYBRID. In horticulture, the production of a new breed by crossing two existing breeds. In robotics, a robot that is part pick and place and part servo controlled, or has the same abilities.

HYDRAULIC CYLINDER. A piston in a cylindrical housing, that is moved by pressurized oil.

INCREMENTAL. Movement broken up into very small pieces, and then taken one at a time.

INTERPOLATION. The process of automatically selecting a path in space based on the positions of the end points of the path; can be used for circular or linear paths.

INVOLUTE TOOTH GEAR. A circular gear having teeth shaped so as to cause lower friction.

LED. Light Emitting Diode. Often used in calculator displays, etc.

LERT. A classification system for robots based on the movements Linear, Extensional, Rotation, and Twist.

LOOK-UP TABLE. In computers, an electronic memory that contains information in table form.

MACHINE LANGUAGE. The code, in 1's and 0's, of electronic logic that is the direct language of a computer.

MASS PRODUCED. Made in great quantity or by methods making each unit the same.

MICROPROCESSOR. A computer contained on a small chip. Developed in the early 1970s, these components greatly reduced the size of computers and robot controls.

MONTAGE. A collection of dissimilar parts.

MULTIPLEXER. The portion of a computer that allows communication with a large number of outside lines, using only a few lines.

OSHA. Occupational Safety and Health Administration.

PALLET. A flat bed, usually wood, used for storage and transporation of goods.

PALLETIZING. The process of placing parts in different positions on a pallet.

PARAMETER. A set of constant factors applying to a particular situation.

PICK AND PLACE. A simple category of robot used to pick parts up and place them down somewhere else.

PILOT PLANT. A full-scale working factory used as a model of manufacturing techniques for other plants.

PNEUMATIC. Of air.

PRESS LOADING. The placing of material in a stamping or other forming machine.

PROGRAMMING LANGUAGE. Any of the computer instruction codes that simplifies programming for people.

PROXIMITY SWITCH. An electrical device that signals when an object is close to it.

QUALITY CIRCLES. A committee that reviews the quality of a product and makes recommendations for its improvement.

RAM. Random access memory. The computer selects the location in which a particular byte is to be stored.

RECOMBINATE DNA. The process of breaking genes apart and putting them back together in a different order.

RELIEF VALVES. Valves which allow excess pressure to be vented.

RESOLVER. An electrical device used to measure the position of a shaft as it rotates.

ROI. In economics, return on investment; profit from this particular source.

R/W MEMORY. Read and write memory. The memory portion of a computer used by the operator to store programs.

SERVO CONTROLLED. Controlled by some type of servo component.

SERVO LAG. The distance a robot arm will move after it has been commanded to stop.

SERVO MOTOR. Any type of motor that has controllable speed or position.

SERVO VALVE. A valve used to control the speed or position of another component by changing its flow.

STRI. A classification system for robot controllers, based on the type of information input: Sequential, Tracking, Registration, Informational.

SOLENOID. An electrical coil with an iron section inside that will pull or push when current goes through the coil.

SONAR. The bouncing of sound waves and the measurement of their echo used to calculate distance.

SPLINE. General term used to describe a shaft with teeth cut on it parallel to the centerline of the shaft.

SPOT WELDING. The process of joining metal together by clamping the pieces and then applying electric current to the clamp.

SPRING STROKE. The length a spring takes when uncompressed.

STEPPER MOTOR. A special electric motor that will rotate in steps of exact size.

TACHOMETER. A device used to measure the speed of a rotating shaft.

TARGET (TV CAMERA). The point at which a light source in a TV camera is aimed.

TCP. Tool center point. An artificial point in space relative to the robot arm where the center of end-of-arm tooling will be.

TRACK SYSTEM. A set of rails on which a robot can travel from one place to another.

TRIGGER POINT. The exact point at which a component will change from one state to another.

VALVE. A device for regulating the flow of fluid into something else.

VLSI (VERY LARGE SCALE INTEGRATION). The creation of a large number of components on a single chip.

WORK CELL. A group of machines all working together on a common part and physically located together.

WORK ENVELOPE. The area in which a robot can work, the limits to its motions.

WORK IN PROGRESS. An accounting term used to express the value of material taken up continuously by the work process.

WORK STATION. A stationary position where work is performed, the work piece moving to the station.

XYZ COORDINATES. A reference to the most common names given to the lines forming a Cartesian solid.

# INDEX